The Viking Mist

Book Two: The Two Realms Trilogy

By: Ariella Moon

THE VIKING MIST

Book Two: The Two Realms Trilogy

By Ariella Moon

Published by Star Tribe Publishing

Editor: Barbara Millman Cole

Cover Design: AM Design Studios

Formatting: AquaZebra.com

ISBN 978-0-9970554-3-6

E-book ISBN 978-0-9970554-2-9

Also by Ariella Moon

The Two Realms Trilogy

The Beltane Escape: Book 1

The Amber Elixir: A Two Realms Novella

The Viking Mist: Book 2

The Teen Wytche Saga

Spell Check

Spell Struck

Spell Fire

Spell For Sophia

Short Stories

"Covert Hearts"

My tremendous gratitude to

Carolyn Cole whose passion for the Two Realms characters and persuasive arguments on their behalf were the magic dust that lit my literary path.

And to Barbara Millman Cole whose steadfast patience, excellent advice, and Socratic edits saved me every time I was struck with Stray Sod.

Prologue

795 CE
I-shona

Three days afore the Norse raid, the sky seemed a promising shade of blue as Disa and her younger sister, Bree, trod the long path towards the island's remote southern shore. Disa indulged Bree's frequent stops to twirl or exclaim over a tiny fairy she swore flew from a clump of toadflax to the top of the monastery wall. Bree did not share Disa's affront at the deep ditch bordered by a high earthen bank, which encircled the monastery grounds. But then Bree believed she was welcome everywhere, and man-made borders were not meant to exclude her.

The monks had no use for girls, favoring the few boys on the isle to help with the pigs, sheep, and cattle. Some of the older boys Disa knew at a glance had joined the brotherhood to learn stone carving or manuscript illustration, trades that chronicled the monks' lore, chiseling in stories of their religion and blotting out the ageless wisdom of the old. She was certain none of them had ever seen a fairy.

"Look!" Bree sang out while abruptly changing course and heading west. She darted ahead, following the path through summer crops, to the fairy mound. "'Tis Muirgen!"

The tabby cat stood up, performed a deep stretch, then sauntered from the *sìthean* to rub against Bree's little legs afore arching into her small palm.

"How do you know its name is Muirgen?" Disa asked when she had caught up with her.

"Because he told me." Bree kissed the tabby's head.

"How clever of him." Disa looked down at the cat's swirled black stripes. The tabby tilted up its chin. Its normally ordinary eyes now appeared overly large and round and gleamed with secrets and mischief.

Startled, Disa clasped Bree's hand. "Away with us. You can visit with Muirgen when we return."

Bree puffed her lips into a pout that quickly vanished. "I'll bring back beautiful mermaid tears for your mound," she promised the cat.

"This is not the way to the mermaid tears," Disa reminded her, eyeing the fertile moor, flat save for the fairy mound.

"We shall see." Eyes glinting, nay doubt from imagined adventure, Bree tugged her forward.

They strolled beneath three yews, the last remnants, according to their mother, of a druid grove from centuries afore the Christians overtook the island. Once past the copse, Bree skipped ahead to search for flower fairies among the yellow trefoil and sky-blue speedwell. When she startled a nesting corncrake among the bright buttercups and summer grasses grown for hay, Disa shooed her back onto the tamped earth path.

As they neared the coast, the trail curved northward, the land grew hilly, punctuated with barren rocky outcrops. Disa spotted a white-tailed sea eagle and heard the faint splash as the majestic bird plucked a fish from the blue-green water. Bree, making up a rhyming song as she skipped ahead, appeared to not notice.

The sun shined high overhead by the time they reached the secluded storm beach covered in small, colorful stones. Disa let go of her sister's hand when Bree pulled loose to dance aimlessly and carefree across the wide rocky expanse.

Watching gulls scavenge for morsels among the rocks, Disa noticed a sun-splashed glint of pale green. Pleasantly surprised, she searched for the translucent yellowish-green stones Bree adored. Almost immediately she plucked three from the craggy ground. Appraising them, she was certain they were not, as Bree believed, mermaid tears or, as Disa had once fancied, tossed upon the shore by sea fairies. Yet, with the pebbles gently pulsing against her palm, Disa could not help but believe they held some sort of magic. Mayhap they warded off monks who forsook the ancient gods and goddesses and disregarded strong-minded, magical girls.

"Are you going to hide more mermaid tears along the monastery wall?" Bree called out.

Alarmed by how far away Bree sounded, Disa looked up. She spied her sister twirling atop a boulder close to the crashing waves. Her hands became boneless and loosed the stones she had gathered. Three gleaming tears clattered on the rocks at her now immovable feet.

"Have care, dear one!"

Mid twirl, Bree halted. Clearly dizzy, she staggered.

"Bree!" Disa ran towards her as a high-crashing wave knocked the child off the boulder. Water engulfed the girl's body, drowning her shriek. Reaching the spot where Bree had fallen, Disa saw only sea foam and water muddied with sand clawed up by the wave. "By the gods, where is she?"

A sudden invisible yank pulled at Disa's thundering heart, drawing her to the left. Mindless of her own peril, she dashed into the rapidly receding wave. Her pulse bounced when a flash of familiar fabric momentarily breeched the surf. Fighting the current, Disa caught hold of her sister's mantle. The sea, like a dog tugging on a doll, pulled Bree backwards. But Disa held tight as Bree, eyes wide with terror, stared up at her through the water. In one swift motion, Disa reached forward, grasped her sister's flailing arm, and hauled her to the surface.

With water rushing about their legs, dragging the sand from beneath their feet, Disa dug in her heels and wrapped her arms high around Bree's waist. For a heart-stopping moment, Bree collapsed—lifeless. Disa squeezed harder. "Breathe, *please*!" Finally, Bree opened her mouth and gasped for air, a strangled, watery sound. Buoyed by this, Disa used all her strength to pull her sister up onto the rocky shore, herself unable to draw in air 'til Bree coughed up water.

When color returned to the little girl's cheeks, Disa crumpled to the ground and held her sister in a tight embrace. "Promise you will never venture so close to the sea again," Disa implored as she rocked the soaked child in her arms.

Exhaustion and shock must have kept Bree from speaking at first, so Disa eased her hold. Like seaweed wrapped around a boulder, Bree's long blond tresses, darkened by the water, clung to her head and neck. Seawater dripped from her hair and clothes, spotting the stones surrounding them. At last, in a small yet earnest voice, she said, "I am sorry, Disa."

"Nay, 'tis I who am sorry. I should have watched you more carefully."

"The stones beckoned you," Bree said, gasping, her voice scoured raw by the sea.

Concerned by the faint blue cast to Bree's skin, Disa said, "Next time, I will not heed their call. I promise."

"Disavow your promise!" Bree's eyes rounded. "Muirgen says magic cannot be denied."

"And Father says Fate cannot be denied, yet I saved you."

"The Fates will be angry!" Bree sobbed. "They will punish us."

"Nay, the Fates intervened on your behalf." *If not, then their anger will be directed at me, not you.* Noting Bree's continued alarmed expression, Disa gently swept her sister's soaked hair off her forehead and declared, "As for the stones beckoning me, I withdraw my vow."

"Thank you." The child slumped in apparent relief, then whimpered, "my foot hurts," as she bent her right knee and drew her leg close. "And I lost my shoes." Her anguished, high-pitched voice sounded so distraught, it panged Disa to realize she had allowed her sister to suffer such a near-fatal fall.

Carefully, Disa brushed wet sand from Bree's foot, revealing a bloody gash that already appeared swollen and bruised. "Can you walk?" Disa asked, even though she knew the answer.

"I do not think so." Bree's pale purplish lips trembled as her teeth chattered.

Dismayed, Disa glanced back to where the path spilled out onto the shore. *Deserted.* Rarely did anyone come to this stretch of coast, let alone know of its existence. She had only stumbled upon it one day following a then four-year-old Bree when that crafty cat had led her down the winding, ill-kempt path. Still, there was a small chance someone might find them. But how long could she wait for such a miracle? Even so, with Bree unable to walk, they would have to stay put for a while.

"You are cold." Disa rubbed her sister's hands and legs to warm them. She wrung water from Bree's léine and mantle as best she could. Finally, she wrapped her own mantle around her little sister, then sat beside her on the cold stones and held her close.

Bree shivered against her.

****

Disa glanced at the headland. The stone people, their faces visible in the moss-green and mud-brown cliff, seemed to stare back at her with disapproval. She dropped her gaze to the rocky beach. Up close, she noticed more colors in the stones. Gull white. Dove gray. Amber. Along with all the shades of blue found in the distant mountains, immense ocean, and endless sky. But when she stared across the thousands of rocks and pebbles that blanketed the shore between her and the headland path, the stones blurred together into white nothingness—the expanse too distant and desolate to discern its variant beauty.

Father may still be ferrying pilgrims to and from the island. And we told Mother we would walk to the southern shore, not head west, then north. Nay soul will search for us here.

From the sea-splattered bed of pebbles beneath them, Disa plucked a round, rain cloud-gray stone and held it up for Bree. "Do you see the white rings in this rock?"

Peering at the stone with great intensity, Bree sniffed, then nodded.

"This is a wishing stone. If we make a wish and throw it, our wish will spread out into the world like ripples in a pond."

"We should wish for someone to come and aid us," Bree said in a weak voice.

"Our thoughts are aligned, dear one." Disa stood, drew back her arm, then shouted, "My sister and I wish for help!" She threw the rock towards the sea. Expectation welled within her as she looked down and saw the wonder in Bree's eyes.

Bree watched the pebble's flight. But when it splashed in the shallows, she turned to her elder sister and said, "It did not go far."

"It went far enough," Disa assured her as her mind turned to more pressing matters.

Looking out over the water, Disa frowned at the overcast sky and heard a distant rumble rolling in behind now frothy waves. *A gale will be upon us soon.* Rushing to shore, the waves seemed to clamber over one another, climbing higher and higher 'til the turbulence formed itself into what appeared to be nine sea nymphs scrambling to gain the shallows where the pebble yet lay, as though scrabbling at a prize. But just as quick, the forms dissolved, empty handed, swallowed in swells that washed over the stone afore slipping back to a darkening sea. Shaken by the illusion, she shivered. *We can wait nay longer.*

"Help will find us on the path if we start walking." Aware of the sand and salt chafing her feet, Disa reluctantly removed her sodden leather boots and handed them to Bree. "You carry my shoes, and I will carry you."

When Bree took the shoes, she dropped a stone into Disa's hand. "For you."

"'Tis beautiful," Disa said, admiring the unusual heart-shaped yellowish-green rock. Then realizing Bree's hands had been empty seconds afore, she asked, "When did you find this?"

"Just now. I called it to me." Bree wiped tears and saltwater from her eyes. "With my thoughts."

A chill tingled down the back of Disa's head, then crawled down her nape. "With your thoughts?"

"Já," Bree said, her tone matter of fact.

"Then 'tis a most special gift, and I thank you." Disa pocketed the treasure, pushing aside her worry whether Bree had injured her head during the fall or too much time on the fairy mound had given her sister delusions of magic. "Let us begin our journey." She knelt so Bree could climb upon her back.

Stones bit into Disa's feet as she staggered across the rocky shore to the pathway home. By the time they reached the sheltering yews, briny gusts rustled the branches. Menacing clouds blocked the sun, raising goose flesh on Disa's arms and legs. She felt Bree trembling against her back. Small for her six years, Bree usually weighed little more than an apple blossom floating on the breeze. But wet to the skin in her sopping clothes, she weighed much more. Carrying her such a great distance, Disa grew parched, and her legs wobbled. The path felt endless.

After what seemed like hours, Disa spotted the fairy mound ahead. *We are better than half the way home.*

"The sky resembles gloaming," Bree observed.

Disa glanced up. *True.* 'Twas nearly twilight.

"What if the snatcher fairies think it *is* gloaming?" Bree's voice rose. "What if snatchers take to the sky?"

"They won't," Disa assured her. "Remember? Just afore gloaming, the sky is streaked with pretty colors. We will be home long afore nightfall."

"But you are limping."

"Someone from the village or the monastery will see us and offer help." *I hope.* "Still worried?"

"Nay." Bree patted her shoulder. "You will protect me."

"Always," Disa promised, even though her legs ached and the cuts on her feet stung. She glanced at the fairy mound, half expecting to see the tabby, then stumbled. But inches afore her knee hit the ground, an invisible force caught her and pulled her to her feet. Her legs buzzed with magic.

Bree gasped and tightened her arms around Disa's neck.

An imposing figure materialized ahead on the path, blocking their way.

Squeezing harder, Bree pressed her lips to Disa's ear. "Who is that?"

"I do not know." Disa heaved Bree higher.

"She's glowing." Bree whispered. "Do you think she is a fairy?"

"She has nay wings."

"Not all fairies have wings," Bree argued. "Muirgen has none."

"Nay fairy could wear such armor and hold an iron-rimmed shield." Disa stared at the nearest yew, wondering if she and Bree should flee to hide among its branches.

"Mayhap she is a goddess who heard our wish."

The woman warrior beckoned them to approach.

"Are we in trouble?" Bree asked.

"I hope not."

The radiant being eyed them. "I bring a message for you."

"You do?" Nervous awe squeaked Bree's voice.

With her elbows locked under Bree's knees, Disa cautiously approached the towering apparition. As best she could, she mimicked an adult tone. "I believe you have mistaken us for someone else."

"Nay mistake." The glowing warrior brushed a lock of sable hair from her eyes. "I am Hlín, and I bear a message for the daughters of Knud from Frigg, Queen of Asgard."

Disa gaped. *The Queen, wife of Odin, knows our father?*

With a single stride, the stranger closed the gap between them, enveloping the two girls in her cool, northern light. Disa squinted against the blue-tinged brightness as she felt the storm winds calm within the woman's incandescent aura.

Hlín lowered her shield and leaned it against her armored knee. "Do you know of the *fylgja*?"

Both girls shook their heads.

"My queen thought your knowledge would be thus. Tell me, what animal comes to you in your dreams?"

The image of a giant brown bear leapt into Disa's mind. Then she remembered her father's recurring dream of a wild speckled horse.

"A frost-white cat with tufted ears and pale-green eyes," Bree declared. "And when I was beneath the waves, he stared at me from atop the rock I had danced upon. But Disa grabbed me afore I could greet him."

"May you remain so unmet," the stranger said, her voice as hushed as a snow-shrouded forest. "The spirit animal you dream of is your soul's protector, your *fylgja*. Beware, for your soul's guardian will materialize to attend you at the time of your death."

Death? Disa glanced up. Muirgen appeared atop the mound and yowled. His ochre ears twitched in the wind. Beyond him, gold and pink threaded the western sky. *Nay snatcher fairies in sight… Yet.*

"Show me your palm," Hlín commanded.

Disa rotated her hand. The tall warrior bent to press a carved birch rune and a glowing golden pouch into Disa's small palm. When Disa closed her fingers around the gifts, the contents of the pouch rustled like dried herbs. Unexpected warmth bloomed across her skin.

Hlín pointed to Bree's swollen foot and told Disa, "When you return home, make certain your sister soaks her injured foot in clean water mixed with the contents of this pouch. The rune is for you. Keep it on you always." She looked directly into Disa's eyes. "Do you understand?"

Disa nodded. "But why tell us about our *fylgja*?" *And death.*

"'Tis not your death time."

"Do you know when we will die?" Disa widened her eyes at the idea.

Hlín held her gaze. "Bree's foot must heal quickly so she will be able to run."

Run? Disa gulped down the heavy fear constricting her throat.

The cool radiance around Hlín shimmered as she stepped back, leaving the girls in the faded light of gloaming. Once more, Disa felt the winds racing from the shore rush about them. Then the resplendent apparition vanished.

Eventide's darkness engulfed them. Two cat paws pressed against Disa's leg, compelling her to stare down at the tabby. Muirgen, his ears flattened to his skull and his worried expression appearing more human than feline, emitted a rumbling warning.

"Muirgen says we should hurry," Bree whispered.

Heedless of the cuts on her feet and Bree's weight on her back, Disa fled towards home as if all the snatcher fairies of I-shona had taken to the sky.

Chapter One
The Guardian

Nine Years Later
The Unmarked Sea

Sullen skies darkened as thunder rumbled in the distance like a war drum, low and ominous. *A storm is brewing.* Disa glanced down at the wet hull of her small craft and gauged the rough water through which she now rowed. *I should turn back.* But instead, she stilled her oars and scrutinized the ring of islands afore her. She glided her gaze past the flat islets that barely rose above the sea, to the towering, sheer-sided islands whose multiple shard-like points stabbed the sky. Not spotting the island of her childhood memory, the distinctive formation harboring the seer's cave, an ache of disappointment closed around her chest. She slumped, resigned to give up and row back to the distant shore, when her heart jolted.

Barely visible in the mist-dimmed light, and hunched in the depression between two rocky peaks, nestled the blackest of black caves. *'Tis as I remember,* Disa thought when she spied the long, narrow boulder wedged into the cave's roof; its distinctive downward point, as pale and sharp as a dragon's fang, impaled the darkness just inside the hollow's inky maw.

Beneath the stone tooth, two glowing amber eyes popped open, piercing the impenetrable gloom.

Disa stiffened and held her breath.

The disembodied orbs scanned the calm sea within the island ring. Then, as if sensing her human presence, the predatory eyes narrowed and searched beyond the inner waters, halting to fasten their malevolent gaze upon her.

In the fixedness of that stare, the monster and years faded, and as though bewitched, Disa saw instead the half mad glare of the Norse slaver and the smear of blood across the man's knuckles when he grabbed her wrist and little Bree's arm. Fear stole the strength from Disa's limbs, and the oars clattered at her feet as a silent scream rose from her core.

A gull's cry brought Disa back to the present. She blinked, startled to find herself at sea. Her hands shook as she gathered up the oars. *Are the gods testing me?*

Inhaling deeply, Disa again peered into the inky blackness of the seer's cave. *Nay eyes?* Her heart pummeled her chest, a trapped bear raging against its fate. Then once more, she felt herself whisked into her past to when she was but nine winters old, newly orphaned, mute with grief, and thin from terror.

"Nay tears," her master, Grambolyn the dwarf, commanded as he pushed her into the cave beneath the protruding boulder. "We will soon discover if ye are the chosen one."

Chosen for what? *Nay thrall nor ambátt had ever met a good end. Disa peered into the blackness.* Why did this dwarf spend his silver on me—only me? Bree needs my protection!

"Go on," Grambolyn insisted.

Young Disa shuffled forward into the darkness. Whatever lurked ahead could not be worse than the thought she might be spared while her sister may already be dead.

The memory enclosed her like stones added to a wall, darkening 'til the last one sealed off all light. She woke to the present pallor of the open sea.

Filled with these unbidden visions of her past visit to the cave, she now wondered, *Had walking beneath the boulder been a test? Had the monster been there when I was nine, but I could not see it?*

With renewed vigor, Disa steered the small boat behind the island. Her muscles burned as she sliced the oars through the choppy water.

The guardian's eyes must be set into a head. The head must be connected to a body. Mayhap the creature is as black as the cave, and that is why I could not distinguish its form. Her thoughts flew like flaming siege arrows.

What weapons might the monster possess? Claws? Talons? Stingers? Scales as hard as armor? A guardian with four legs could kick and trample. One with wings could assault me from the air. Or a serpent could—

A thunderclap, far louder and closer than the last, boomed around her. Disa flinched. *I must focus.* She dragged the oars through a cross wave that rocked the boat.

Approaching the far side of the island, Disa glanced back but could no longer see the cave or its guardian. *Good, the monster cannot see me either. I am safe...for the moment.*

The windward side of the twin hills slanted steeply, then disappeared into the sea. *Nay cave opening on this flank. If I scale this backside to the seer's cave, then jump down to the cave opening, I might slay the guardian afore it can react.* Her knuckles, raw from the cold and rain, whitened as she tightened her grip on the oars. *She who dares, wins.*

Still, unease gnawed her stomach.

Disa unclenched her shoulders beneath the circular shield slung across her back and skimmed her gaze along the length of the island. *Nay place to moor the boat.* Her shoulders knotted again. She searched with greater care. *There!*

When twice she cast the landing rope and failed to snag the one boulder that hadn't been smoothed into the hillside by the wind and sea, Disa cursed. Then, though she doubted it would improve her aim, she lightly brushed her hand across the enchanted moonstone embedded in the iron hilt of her sword.

A warm tingle twitched her hand.

On her third try as waves swelled around her, she hooked the rope over the protrusion, then pulled the small vessel alongside the rocks. Fat raindrops plastered Disa's plaited hair to her nape and pinged against the iron rim of her wooden shield.

"Dear Goddess!" Disa swore beneath her breath when she nearly fell into the sea while clambering out of the tipsy boat. "I am soaked enough!" With sharp pains biting her shoulders, she heaved the sodden craft upon the rain-slick boulders. Hopefully, the slap of waves and cries of seabirds would muffle the scrape of wood against the gray rocks.

Neither Grambolyn nor the dwarf's mage warned me of the cave guardian. If this is another test... The thought set her grinding her teeth against each other.

Fueled with fury and a sense of urgency, Disa eyed the impossibly steep hillside through eyelashes jeweled with raindrops. *Rocks, heather, sedge, and mud.* She braced for a slippery climb.

A splash sounded. Lightning-fast, Disa whirled, ready to unsheathe her sword. But her fear allayed when she saw a puppy-eyed seal surface between the waves, only its slick head visible. Her mind flashed back to a wide-eyed Bree, staring up at her through treacherous waves. When the seal dove beneath the blue-gray water, Disa fortified her resolve with a lungful of air, then turned to begin her climb.

"Goddess, protect me," she whispered.

Already fatigued from rowing, Disa's muscles quivered as she scaled the sheer hillside. When the boulders were too far apart, she clawed finger holds into the mud and sedge. *Pray the heather does not give way beneath my feet!*

Over the thunder and rain, Disa heard a second splash far below. Recalling the seal pup and straining to grasp the next handhold, she paid the sound little heed.

Loamy earth smells filled her nostrils as she continued her ascent. Rain needled through her cloak, dampening her padded tunic and close-fitting trousers. She reached across mud-splattering rivulets that carved narrow vertical paths in their rush to the sea. In shoes grown heavy with mud, she attained the summit, pulled herself up over the cliff edge, and rose to her feet atop the wet boulder.

Then she slipped. Her stomach whooshed as, strong and quick, she regained her footing. But in the act, she dislodged several rocks that slid over the front ledge.

With dismay, Disa heard them ping against the top of the boulder wedged into the cave roof. As she peered over the edge, she saw three of the larger stones clatter, roll, then continue their downward trajectory, tumbling to the ground in front of the cave opening.

Distraught the element of surprise had been lost, Disa pulled back. She silently eased her sword free of her waist scabbard and pressed her thumb to the moonstone inlay. With an icy burn, mage magic surged up her arm. The sudden bitter taste coating her tongue made her shudder. Magic warmed her core, pumping her blood so hard, she felt her veins bulge. Knowing her increased strength and focus would not last long, she prepared to jump.

Then Disa noticed the sudden silence, the absence of seabirds and thunder. Even the persistent rain had stopped. A tingle tip-toed across her shoulders, and the back of Disa's head prickled. Sensing something enormous slithering up the hill behind her, she hefted her sword with both hands and spun.

Whatever was advancing upon her increased its speed. Heather snapped. More dislodged rocks splashed into the sea. Disa cautiously walked back across the boulder and peered over the rear ledge to see what danger approached.

The cave guardian!

Below her, in the storm-gray light, two distant glowing orbs pierced the mist. The monstrous winged serpent slithered at rapid pace towards her. Seawater glistened on the creature's dragon head and serpentine body. As it neared the crest, the beast halted to ruffle its leathery wings, shaking off great beads of water. One wing stuck out at an odd angle as though it had been broken but had not mended properly. Lighter now, its pace quickened, and as it slithered closer, Disa could see the feral glint shine in the cave guardian's amber eyes.

Disa tightened her grip on her sword. With a sweeping glance, she assessed her foe. Nay horns. *Nay legs nor talons. 'Tis not breathing fire…yet.* She frowned at the thrashing barbed tail. *Could that be poisonous?* The shield-maiden stiffened her stance and positioned her weapon.

When the guardian rose afore her, it lowered its jaw, exposing long, jagged teeth. Disa recoiled as the monster's rancid breath gusted against her.

Then aiming to behead, she swung.

Time slowed.

The serpent dodged her blade.

Disa sidestepped and jabbed, upper cutting at the last moment when the beast thrust to bite. Her sword connected with the monster's jaw and bounced off, but not afore Disa heard a *crack.*

The guardian shrieked, flailed its wings, then coiled.

Time sped up.

Hissing, the monster flattened its mangled wings and slithered sideways. In its haste, the guardian knocked against a large boulder half embedded in the hillside and sent it tumbling. Thrown off balance, the creature slipped downhill after it. Disa hoped the monster would fall back into the sea. Instead, it slowed its fall by sidewinding its body and whipping its barbed tail. Losing ground, the creature swept its wings wide, then awkwardly pulled them close in a bat-like movement that halted the beast's downward slide.

She cursed, then swiveled towards the cove, ran to the edge of the boulder, and leapt. Disa landed hard on the slippery cave roof. Pain spiked up through her feet and into her legs. Still clutching her sword, she ran two steps, swung out her arms, then leapt again to the narrow beach that fronted the cave mouth.

A deep, curved trough scored in the sand marked where the guardian had slithered from its lair. Disa hurried inside the inky cave. Her leather shoes connected with something hard yet brittle scattered on the ground. *Sticks?* When her eyes adjusted to the darkness, she gasped. *Broken bones! Dear Goddess, had the guardian eaten the seer?* She peered more closely. *Sea birds and seal bones, not human remains.* With a long exhale, she released the terror trapped in her throat.

With magic still burning her veins, Disa scaled a tall pile of boulders off to the right, just inside the entrance. Thus ensconced, she waited, braced for battle.

Discharging an inhuman screech, rivaled by the alarmed flailing of its wings, the cave guardian thudded onto the beach with a jarring *thump, thump.*

Hidden in the dark, Disa shrugged off her shield and pressed it to her chest as she squeezed sideways into a narrow gap between the highest boulder and the cave wall. Her heartbeat filled her ears as she adjusted her grip on the dwarf-forged sword. A growing heaviness in her limbs warned her the mage magic would soon be spent. Disa stilled.

The guardian slithered into the cave, its ponderous serpent's body snapping strewn bones. It glided past the fang's tip protruding through the roof. For a moment, Disa thought her foe might disappear into the cave's black depths. Instead, it stopped. Amber beams cast by the creature's glowing eyes swept the chamber as it turned towards her.

Disa shrank farther back in the crevice and raised her sword.

Slow and deliberate, the monster coiled, then rose 'til its head was level with Disa's hiding place. Tawny light slid into the cranny, half blinding her. Penetrating the eerie quiet, the cave guardian flicked its long tongue. Then, in the imperceptible second between one heartbeat and the next, the serpent struck.

Its blow missed Disa but knocked the front boulder askew when the monster's teeth crunched against the stone. The huge rock scraped and twisted sideways, halting the beast's assault, but threatening to entomb her. Disa squeezed around the precarious boulder then thrust her sword into the dragon's snout. Greenish blood spurted her cheek and shoulder.

With an angry hiss, the guardian reared back. Disa seized the moment, summoned her enhanced strength, and in a counter move, pulled her weapon free. The beast swiveled aside and shook blood from its nose.

Disa darted forward and hurled her shield. Its iron-rimmed edge smacked the guardian's left temple, then the shield thudded to the ground as Disa leapt for the monster's throat and drove in her sword.

Eyes flashing, the cave guardian screeched, whipping its serpentine body and tail. Fearing the barb, Disa pressed her knees to the creature's convulsing throat and struggled to free her sword. The dragon serpent flapped its leathery wings, swooshing the musty cave air. But then the beast pulled in its wings and stilled itself. Breathless, Disa waited. With sudden speed, the monster lashed its barbed tail towards Disa's heart.

Unearthly cold breath ghosted against Disa's nape. An enormous brown bear flashed in her mind. *Nay!* "Begone!" she commanded the *fylgja* as she struggled anew to free her sword from the winged serpent's throat. "I have nay intention of dying this day."

The spirit bear roared; its ethereal breath shrouded her like a thunderous cloud.

Again, time slowed.

Cave air thickened, then warped. Nearly invisible waves pressed against Disa, stifling all sound. Within arm's reach, the serpent's tail froze midair.

"Not today!" Disa shouted. To her ears, her words sounded slow and distorted, as though they traveled between worlds.

Her *fylgja* grumbled.

The air cleared. Sound traveled freely again.

Time accelerated.

Released, the barb continued its course, arrowing towards Disa's heart.

With one frantic yank, Disa dislodged her sword from the monster's throat and slashed off the deadly barb.

The cave guardian dissolved into smoke, then vanished.

Winded, Disa crashed to the ground. A tingle spiraled into her belly as she sensed her *fylgja* recede. The magic warding the cave lifted, and a warm glow flared afore her.

Chapter Two
Revelation

Fire crackled within a ring of stones on the ground, casting flickering shadows on the cave's earthen walls. An old woman, her face supernaturally unlined, her hair white and wild, stirred the flames with a charred wooden staff.

"Ragna."

"Grambolyn's thrall. Have the years passed so quickly? I did not expect you so soon." The sybil's fathomless dark eyes rose and met Disa's. "Where is the dwarf? Did he not accompany you?"

"Nay. I came alone." *Did you not foresee my arrival?*

"Pity. The guardian was meant to deter Grambolyn…" Ragna's speech sounded distracted as she scrutinized Disa. "Not you." She circled Disa, her gaze coming to rest upon the bloody weapon Disa still held. "Was it the reek of dwarf-forged magic in your sword that drew my serpent's attack? Pity, indeed."

"Why deter my master?"

"I wearied of him enslaving human children to serve his quest, then discarding them when they proved unworthy."

Discarding? A sudden dizziness threatened to fell her. "I am not the first?"

"Nay, the third. Two small boys came afore you." Ragna swirled the embers and clucked her tongue. "Handsome. Dark haired. Too young to carry out that implacable dwarf's plans."

"What became of them?"

"Grambolyn traded both to a fairy. Exchanged them for information about Mab. Beyond that, I know not their fate."

Traded? Dear goddess!

"There was a fourth child—in case you failed. To save her, I cloaked the cave from Grambolyn's sight, blocked the dwarf from my foresight, and summoned the guardian to vanquish Grambolyn should he ever attempt to enter my sacred space." Ragna looked up. "My apologies. It appears I nearly blocked you as well."

Shaken, Disa seized a cloth drying near the fire and used it to wipe the foul-smelling blood from her face, then sword. Once finished, she set her expression into one of outward calm, then sheathed her sword and retrieved her shield. She slid her hand into a hidden pocket in her tunic and fingered her birch rune; her knuckles brushed the heart-shaped stone Bree had given her. *They are safe.* Hlín's protective symbol clattered against the third hidden object, the tiny crystal sphere given her by the dwarf's mage.

Did the mage know of the other children? She flashed on her last encounter with the unnerving magician, Grambolyn's young daughter. 'Twas the first time she had encountered the baby-faced dwarf away from the child's musty chamber filled with pungent potions and drying herbs. Mayhap it had been a trick of the sunlight and fresh air, but the mage's flint-gray eyes had gleamed light green, and though the dwarf village was many days walk from the nearest shore, the child had smelled like a salty sea breeze.

"Keep this crystal until the end," the mage said, her little girl voice replaced by one inflected with maturity and concern. The young dwarf fleetingly assumed a womanly and regal presence as she pressed

the small orb against Disa's palm. "Or use the crystal when the moon is bright, and you are ready to return."

Afore Disa could thank her, the mage had vanished.

Disa shook her head to clear the puzzling memory. *Remember your training.* She stared past the farseeing crone and searched the dank shadows for danger. *Nothing.*

Bats clinging to exposed roots in the ceiling near the hearth rustled as Disa drew closer. Her throat burned from the smoke mingled with the seer's earthy, dragon's blood scent. Disa untied an indigo pouch from her belt. Then clearing the roughness from her throat, she carefully offered the present to the old woman. "Grambolyn sends his regards."

Ragna glared at the pouch afore lifting her gaze to meet Disa's. "After I warded against Grambolyn, he retaliated by forcing his mage to dull my powers. I know this because the magician sent me a vision. Whether she meant to taunt me or warn me, remains uncertain." Ragna prodded the fire. "Yet I managed to retain some sight, enough vision, within my scope. But I shall accept nay regards from that dwarf, save the restoration of my full abilities."

"Mayhap Grambolyn's intent is to restore your goodwill."

Ragna snorted. "He has anticipated my ire; yet nay mage has materialized to undo his bidding."

Yielding, she set aside her stick, took the sack, and lifted the weighty gift from the soft cloth. "A diviner's crystal!"

"Mined and honed under the Wolf Moon."

"Quite the gift. It sings to me." The seer caressed the orb. "But neither my approval nor my forgiveness shall I grant him in return." She paused, her eyes narrowing, then carefully replaced the crystal inside the pouch. "Mayhap his young mage has magicked this sphere to spy on us."

Disa stiffened. *My master would not spy on me.* Filled with sudden doubt, she thought of the crystal hidden in her pocket.

The seer left the fire to store the sphere on a low stone-lined shelf cut into the earthen wall. When Ragna returned, she carried a three-legged stool and placed it near its twin by the fire. "My grievance is not with you. Rest from your long journey."

Afore lowering herself onto the hard seat, Disa positioned the stool so she could heed the entrance.

The diviner handed Disa a drinking horn, then watched as Disa quaffed the mead. "Grambolyn trusted you to journey alone and to carry a gift many would kill for."

"I have never given him reason to distrust me."

"Hmm." Ragna studied her. "I can nay longer see the frightened nine-year-old you once were. You dress as a male warrior in tunic, trousers, and padded leather." As though affirming her appraisal, she concluded, "Grambolyn has transformed you into a shield-maiden."

"I transformed myself." *That I might be like Hlín, and never again be so vulnerable.* A familiar ache clouded Disa's heart as she remembered the harrowing voyage in the slave trader's boat with Bree. Still fresh in her mind was how she had held her little sister's hand as the slave ship docked in the boisterous Dubh Linn port. A Norseman, his face and neck covered in markings, had then divided the captured men and boys from the women and girls, prodding them into separate empty animal pens. Disa slipped deeper into the past, recalling that last day.

"Someone is watching us," Bree said in a small voice.

Twisting, Disa caught a male dwarf observing Bree. Mayhap sensing her glare, the dwarf lifted his gaze. Seeing nay compassion or warmth in his eyes, Disa quickly turned back and knelt afore Bree.

"Nod if you remember why Hlín gave us the medicine to heal your foot," she whispered.

Bree nodded.

Disa glanced over her shoulder. The dwarf had advanced within earshot. She whispered directly into Bree's small ear. "They have not shackled you. So the moment you see a chance, you must do exactly as Hlín instructed."

Bree's lower lip quivered. Sniffing back tears, she scanned the throng outside the pen.

Disa stood, dizzy from hunger and fear, knowing Bree could never have the courage to run, not by herself. Sudden pain interrupted her fretting when a slaver grabbed her arm in a crushing grip. As he dragged her towards the auction block, Disa kicked and hit, loudly cursing him in both her parents' native languages to draw the crowd's attention away from Bree. But with shocking speed, the lurking dwarf bought her and spirited her away, affording her nay chance to even glance back, nay chance to ensure—

"Grambolyn would have saved your sister if Fate had allowed." Ragna's commiserative voice wrenched her from her ruminations.

"So I have been told." *Since Bree vanished without me seeing who bought her, I may never learn her fate.* She lowered the horn, memories of being torn from Bree stealing the color from her knuckles.

"Did Grambolyn tell you why he picked you from all the other slaves that day?"

Disa shook her head.

"Because he saw you were quick-thinking, loyal, and self-sacrificing. Dwarfs live long lives. They plot and scheme and wait. But humans… Humans act decisively. He recognized that trait as being strong in you, especially when he observed you put your sister's wellbeing ahead of your own." When the old woman's warm hands enveloped hers, Disa saw the seer's eyes belied her words. At the least, the diviner was not revealing the full reason for Grambolyn's choice.

Disa relinquished the drinking vessel.

As Ragna hung the horn on a peg in the shadows, Disa flashed on the raiders who had captured her family. *Bree–probably sold into slavery; her fate unknown–Father—murdered.* She pressed her palm to her heart, remembering the wild speckled horse that had appeared...after. *Mother*— A brush of the crone's fingers across Disa's forehead vanquished the painful memories to a shadowy corner of Disa's mind.

"When the Last Fairy King and his dwarf advisor, Gudbrand, were murdered, the dwarfs not only lost their prominent, much beloved leader, but also their esteemed access to the Court. Mab reduced them to working in the mines where they could neither spy on her nor create new alliances." Ragna drew in a deep breath then released it. "Which bring us to you. Grambolyn must believe you are ready." The old woman glanced down at the dwarf-crafted sword belted to Disa's waist.

Aye. He armed me.

"Shall we see if his faith is warranted?"

Disa nodded.

The crone glanced around the cave. "What calls to me?" A large, bone-white seashell flew out of a niche cut into the wall and landed in the seer's hand. "Hmm. Interesting," she mumbled as she caught the lap-sized sheepskin that flew off her bed. "Anything else?" Ragna whirled with her arms outstretched. "Nay? Then we shall proceed." She settled on her stool and placed the items on her lap.

"Blood?" Disa asked.

"You remember." Ragna's fathomless eyes conveyed approval. "Three drops."

Her lips set in a determined line, Disa removed the plain, circular brooch from her cape, then pricked her finger with the pin.

With swiftness that belied her age, the old woman clutched Disa's finger, held it over the seashell, and squeezed out the first three drops of blood.

When Ragna's grip slackened, Disa pulled her hand back, placed the brooch on her lap, then pinched her finger to stem the crimson flow.

Ragna shook the seashell over the fire, swirling the blood. "Oh!"

"What do you see?" Disa leaned forward.

"Quick! Take the shell."

Disa grabbed it.

"Hold it steady." Ragna waited 'til the liquid settled. Then with a mere glance at the becalmed blood, she bent over the shaved sheepskin as though it were a smooth canvas. Magic flowed like ink from Ragna's finger, imprinting lines on the pale leather.

While Disa watched with wonder, a map depicting a rocky, mountainous island emerged.

The stench of sulphur and the smell of ancient books began to fill the cave. Disa felt a disturbance in the air and glanced up to see Ragna flicker and a young man wearing white druid robes appear in her place. When Disa gasped, the image vanished.

Then Ragna, though once more her true form, spoke in an otherworldly voice, deeper than her own:

"Many mountains you will see.
Choose from one, or two, or three.
But pick carefully your course,
To avenge the fallen dwarf."

"'The fallen dwarf,'" Disa repeated. "Do you speak of Gudbrand, the ancestor Grambolyn believes was murdered by the Queen of Fairy?"

Ragna's image wavered like the shadows on the wall as the druid reappeared. Disa schooled her features into a neutral mask and memorized the man's appearance—unkempt brown hair, pale skin, narrow shoulders, a pointy, goat-like beard—afore he faded back into Ragna and continued in his bass, sing-song voice:

"On a dark eve filled with hate,
The Fairy King met his fate.
So, too, did Gudbrand the Dwarf,
Murdered by the queen who morphs."

"Morphs?" Disa rose from the stool. Her rash motion caused her brooch to fall from her lap and tumble to the cave floor. Lost in this new idea, and oblivious of the trinket's soft thud or the silence that followed, she inquired, "Into what form does Queen Mab morph?"

"A golden dragonfly, 'tis told."

Disa's hold on the shell tightened. "And what is the queen's greatest fear?"

"The truth," the druid said without hesitation, his deep voice resounding through Ragna.

Truth? Who slayed true king and trusted dwarf? Her pulse raced as she considered the adversities of her quest. "Pray, what allies will I find within these mountains?"

Ragna's possessed form leaned forward and peered at the blood now bubbling in the shell.

"From the future, a powerful, flame-haired woman shall you ally.
And from a time without time, a dear one at last shall you eye.
You will journey to a realm where few humans find vent,
Enticed, then enslaved, they are ensnared in wilderment."

Enslaved. Disa closed her eyes against the sudden rush of

more unwelcome memories. *You were a mere child when the slaver stole you and Bree,* she reminded herself. Deep within her core, she hardened her resolve. Then she realized the otherworldly voice was still speaking.

"…if it be friend you meet.

And your enemies will be difficult to defeat."

Ragna's body fluttered 'til the young man materialized again.

Leveling her gaze, Disa asked, "'A time without time…' Fairy?"

The druid stared blankly.

Disa stiffened her spine. "I must journey to Fairy?" she pressed.

The druid inclined his head, then faded.

Ragna leaned back, herself once more.

"And the young druid?" Disa asked in earnest. "Will he be there as well?"

"Druid? What druid?" Appearing dazed and disoriented, Ragna rubbed her finger against her thigh, as if to quell the magic that had flowed from her fingertip. "Be where?" The map on her lap caught her eye. She flinched, as though startled by the sheepskin's appearance. Then, clearly fascinated with the ink drawing that now adorned the skin, she studied it as though she knew the place well.

"For over three centuries, the dwarfs have searched for proof Mab killed the Last Fairy King and Gudbrand the Dwarf. 'Tis unfair of them to place their burden upon you," Ragna said in an almost absentminded tone as she traced her finger along the mountain trails that now stretched across the length of the sheepskin. She looked up at Disa. "Is this quest the price you must pay to gain your freedom?"

"Taking up the dwarfs' cause was my idea, not Grambolyn's." Disa retrieved her brooch from the floor, and as she fastened it to her cape, said, "I doubt any proof still exists. But I vowed I would search, or, should my efforts prove unsuccessful, I would at least stop Mab from killing again. Grambolyn's offer to free me when I return is an act of gratitude." Done securing the brooch, she looked up and pinned her gaze on Ragna. "Not a condition."

"You truly believe this to be true?"

Disa's faith faltered. "Grambolyn vowed he would search for Bree."

"I see." There was something pitying in Ragna's expression.

Shield-maidens strike fear and awe, not pity. Disa thrust the shell into the seer's hand. The blood sizzled, then the liquid transformed into thin streams of red vapor that rose to the earthen ceiling, scattering the bats.

"I've tried to divine your sister's fate," Ragna confessed, staring at the streams as they dissipated above them. "Nine years ago, after you left with Grambolyn, and more recently when I sensed your need."

Disa clasped her hands together to keep them from trembling.

"Thick mist." Ragna shook her head. "I saw vapors, nothing more. I could not divine their meaning. I am sorry."

Tears pressed upon the back of Disa's eyes as she stared, first at Ragna, then at the shadows bobbing on the cave wall. She cleared her throat and met Ragna's remorseful gaze. "Mayhap we will meet again in another nine years." *Long after my thralldom has ended, and Bree and I are free women.*

Ragna's shoulders sagged as she handed Disa the map. "I fear Fate may not allow it."

Uncertainty speared Disa's thoughts. *Has my fate been revealed, or hers?* Her hand quivered as she reached for her shield. "Have you any further counsel?"

Ragna glanced at the diviner's crystal, then frowned as though the gift conjured unpleasant dealings with Grambolyn.

Shifting, she met Disa's gaze. "Always remember. You were not *born* a slave."

What does that matter now? Disa slung her shield over her back, then tucked the map inside her padded tunic. Her thoughts in turmoil, she turned from the seer, then strode towards the mouth of the cave.

"Beware of losing your heart when you encounter the fay."

The seer's unbidden warning stopped her. Nettled, Disa stiffened and glanced back. "I am not easily beguiled."

Ragna's unfathomable gaze ensnared her. "Then you have never met a fairy."

Chapter Three
Talfryn

Eighteen Years, Nine Months Later
The Realm of Earth, Fairy

Beacon appeared pensive, watchful, his son Talfryn observed as they stood with the other Trooping Fairies inside the Aurelia Antechamber. The amethyst encrusted tunnel afforded an unimpeded side view of the Blessed Court's Great Chamber beneath the hawthorn. Talfryn attended his father, head of the Noble House of Storms, with fixed focus, alert to any instruction the Trooping leader may deem appropriate. He'd never felt more need to be close to his father, for he'd never known a more perilous moment in Fairy.

A murderer lurks among us. The inconceivable thought looped unceasing inside Talfryn's head. He closed his eyes against the memory of the fireball that hours ago had lit the night like dragon fire, torching the forest—not all of it, just the seat of the Noble House of Boughs—leaving the fairy Merrybelle dead. *It could have been Lily who died instead of her sister.* A sick feeling churned his stomach as he remembered how Lily, his closest friend aside from Fenella, the Thaness of Thorburn, had sustained unhealable burns.

Unhealable!

Even in the antechamber beneath the earth, Talfryn's eyes burned. His throat constricted from inhaling smoke laced with the acrid scent of malevolent magic. He longed to draw a deep breath, to inhale the sweet fragrance of summer flowers and the cool mist wafting off the lakes and river. But though his nostrils flared, his lungs refused to expand. The burn of a spent evil spell along with suffocating smoke and ash still coated his nose and throat.

His father spoke in a conspiratorial whisper. "Not since the murders of the Last Fairy King and Gudbrand the Dwarf has Fairy suffered such an unspeakable crime."

"Those unsolved murders were centuries ago," Talfryn said in a hushed voice.

"Aye." Anger tinged Beacon's tone. "Queen Mab never discovered her late husband's killer." He muttered something under his breath that sounded like, "Not that she tried too hard."

"Mab better be more successful this time." Talfryn peered over the heads of the waiting Trooping Fairies lined in front of him. Beyond them, crystal lanterns, conjured by the Noble House of Glint, free floated inside the Great Chamber. The lights cast a warm glow on the Blessed Court gathered below. *The murderer may be one of them.* Then he swept his gaze over the Trooping Fairies. *Or one of us.*

Airy music reached his ears, signaling the arrival inside the chamber of charmed wind chimes representing the Noble House of Lilt.

As if heralded to assume their posts, the crystal lanterns scurried to the center of the ceiling and formed two long rows. Below the lights, fay scattered, leaving a wide path.

Talfryn's focus returned to his fellow Trooping Fairies. All had forsaken their customary resplendent attire for mourning grays. Instead of their usual hair primping and puffing of chests in anticipation of the Grand Procession, the nobles shook their heads and murmured among their factions. Even the usually carefree, merrymaking members of the Noble House of Rings acted subdued and had forsaken the glamoured mica powder they often applied to emulate The Six's shimmering bronze skin.

An enchanted breeze representing Talfryn's family's faction, the Noble House of Storms, swirled into the main chamber, vanquishing the lingering stench from the fire. The gust released a cascade of golden specks. As the flecks touched the ground beneath the lanterns, they formed a narrow, gleaming, gold carpet. Enlivened wind gusts rustled the fairies' wings.

Hypnotic music dove between brownies, hobs, and house elves, then swooped among dwarfs fresh from the crystal mines. The music circled a suspicious gang of goblins who had traveled through a tenacious ground mist and smelled like damp dogs.

Then the melody shifted. Chimes howled, evoking the legend of Queen Mab and the Five Golden Dragonflies. Waiting in the antechamber, Talfryn closed his eyes and imagined The Six caught by a scorching, once-in-a-thousand-years wind, transported across land and sea, then pushed forward centuries in time to Fairy. He wondered why the legend accounted for Mab's bronze skin and dragonfly wings yet failed to explain her unnerving white eyes.

He considered, not for the first time, whether Mab had landed in the wrong Court.

If only Queen Titania had not died in childbirth, and the Royal Heir had not perished as well.

Sudden gasps and sympathetic murmurs interrupted his reverie. As Talfryn opened his eyes, the sickening scent of burnt flesh and scorched wings assaulted him. He quickly located the source. Two factions ahead, Lily's father's family from the Noble House of Ruse welcomed Lily and her parents, three hunched shadows, into the procession line.

"Merrybelle just died!" Talfryn exclaimed to his father. "Why would Mab force the Noble House of Boughs to attend?"

"Why indeed?" Beacon's words stretched out, as if he considered the matter worthy of investigation.

Talfryn broke from the line, but Beacon clasped his shoulder and hauled him back. "There is no time. Speak with her later."

"But–"

As if sensing the commotion, Lily glanced back. The tense set of her jaw and the unnatural shine in her tilted lake-blue eyes conveyed her grief and pain.

Talfryn's stomach tightened. "She didn't conjure a glamour to hide her oozing welts or singed wings." He bowed his head in a silent salute. "She's reminding everyone there is a murderer in our midst," he said in a low voice. Then with fierce conviction, he added, "May the evil spell-caster be exposed afore he strikes again."

"Aye." Beacon narrowed his eyes. "Afore he, or she, *kills* again."

I hope whoever committed this unspeakable crime suffers a slow, painful, Cold Iron death.

Nettle, Talfryn's squashy-faced pixie, scrambled out of Talfryn's cloak pocket and crouched on the Halfling's shoulder.

Talfryn batted at the pixie's hand. "Quit tugging my ear."

"Fine! But a flower fairy told me Merrybelle had caught the Prince Consort's fancy."

"Curious." Beacon leaned close and whispered, "Never speak of this in public again."

Nettle gulped.

Swelling, pompous notes spilled from the Great Chamber into the anteroom. An elder at the front of the procession raised the banner signifying the Seven Noble Houses. The nobles trooped into the cavern, treading the air inches above the gold carpet.

Lily's mother faltered, her foot marring the gold-specked ground. Her husband caught her arm, preventing a tumble, but a hill fairy in the crowd cackled.

Talfryn, who had hunched his shoulders and kept his chin down, straightened to his full height and glared. The movement drew attention away from Lily's parents. Dozens glowered, judging him, the Halfling who dared walk with pureblood nobility.

Once all the Trooping Fairies were inside and had joined the masses, Talfryn lost sight of Lily. He shouldered his way to the shadowy back wall where he would not block anyone's view. A tiny thistle fairy flew past his ear and landed on a parasol mushroom growing out of the wall as the chimes stilled.

"There's Lily," Nettle said.

"Do not point!" Talfryn watched Lily hug her slender arms to her waist. *I will search every crag, meadow, and tree in Fairy 'til I discover who killed your sister and injured you.*

Nettle lowered his arm. "Did Mab call us together because of the fire?"

"She'd better have." He scanned the crowd. "But why summon the Border goblins? They aren't members of the Blessed Court."

"Mayhap *they* torched the forest," Nettle speculated.

With an abrupt disturbance of air, Brindleburr materialized, setting several fairies aflutter. The Prince Consort's midnight-blue tunic and black silk dimijes hung like wilted flower petals on his sleek frame. A dazed expression contorted his flawless features, and he squeezed his eyes shut for a moment afore flicking his long-fingered hands, compelling those closest to the gold runner to step back.

Brindleburr stiffened his neck and snapped up his chin afore striding to the head of the golden path. At his command, roots pushed upwards from the earth, releasing a loamy smell as they intertwined to weave a pedestal. Twisting towards the assembly, he adjusted the gold circlet banded around his head. Without his usual flourish, Brindleburr announced in a monotone, "Mab, our queen."

In a flurry of gilded fairy dust and heat, Mab materialized on the pedestal. Her enormous dragonfly wings thrummed, then stilled. The queen's slaves, three human children, each wearing a binding-spell circlet around their head, appeared behind her. Mab's spicy foreign scents wafted through the gathering, tickling Talfryn's nose. As the queen stared down at the assembly with her eerie, nearly solid white eyes, flower fairies hovering in the front cowered.

Talfryn gave his Norse armband a nervous twist.

"I've called you together," Mab said in her husky voice, "because there is a thief among us."

A thief? What about the fire and the murderer who caused it?

"I thought all hill fairies were thieves," Nettle muttered.

Mab rolled her unnerving eyes towards the pixie, and Nettle shrank back.

"Someone...Stole...From...Me!" Each word stabbed the musty air like a Cold Iron spear.

Fairies gasped and glared at goblins. Goblins clicked their pointed teeth in defiant response.

"The Aeternitas Potion has vanished." Mab's unsettling stare probed the crowd. She clenched her hands, and Talfryn had the odd premonition fear loomed behind the queen's anger.

"Doesn't the potion bestow immortality?" Nettle asked.

"None of us need it!" a particularly mean-faced goblin shouted. "Why would we steal it?"

Mab arched her glittering brows. "Not everyone in Fairy is immortal."

Talfryn's uncertain glance darted to the three human children.

"Do you believe the theft is linked to the fire?" the head of the Noble House of Ruse asked.

"Who can say," Mab answered vaguely.

Lily's parents exchanged a confused glance.

"But you *are* investigating the fire and our daughter's death?" asked Lily's mother, Willow Weep.

"The matter is utmost in my mind," the queen responded.

Talfryn sidled along the wall. Grim-faced, Beacon lobbed him a warning glance, freezing Talfryn in his tracks. Foiled, he hunched, cloaking all but his bright hair in shadow.

"The Aeternitas Potion is given to humans to whom we wish to grant immortality," Mab reminded the crowd.

"So she can extend their enslavement or trade them to the Unblessed Court," Nettle whispered. He inclined his head towards the two raven-haired boys and the redheaded girl.

Talfryn shifted from one foot to the other.

"When their binding-spell circlets fade to white, or the fairy who cast the spell dies, their enslavement ends," Nettle added. "Then they can escape Fairy, unless they drink the potion or are captured by the Unblessed Court."

Talfryn's morning meal of flaxseeds and dandelion blossoms roiled. For the first time, he noticed the humans' spell rings had faded over the years from light-up-the-forest bright green to a dull shade of sage. He wondered how much longer the binding spells would last.

"Most of us avoid humans!" a forest fairy protested.

"Unless we're playing tricks on them," a hill fairy named Rocco said, then half snorted. His searching gaze leapfrogged through the crowd. Several hill fairies nodded, raised their arms, and made a door-knocking motion with their fists in apparent support.

The three human children glanced at each other afore lowering their chins and melting farther back.

Stop it! Talfryn wanted to shout. *The Noble House of Boughs was attacked. Merrybelle died. Lily is injured! Why are we discussing an immortality potion?*

Mab raised one of her elegant hands. The assembly quieted. "There are those among you who have had *significant* contact with humans. Beacon?" Mab fixed her gaze on Talfryn's father. "Didn't your son experience such an encounter?"

The crowd swiveled to face Beacon. As his father approached the queen, the sick feeling in Talfryn's stomach coiled.

Beacon's snowy hair curled over the collar of his spotless thundercloud-gray tunic, mourning hue for the Noble House of Storms. "Talfryn did assist Lady Fenella, the Thaness of Thorburn."

Eyes wide, the ginger-haired slave girl glanced up.

Heads turned towards Talfryn. He glared back.

"As a result, an alliance was formed with Clan Thorburn." Beacon continued with an air of humble pride and greathearted service. "The Blessed Court may now live or pass safely throughout their holdings in the northern kingdom. Clan Thorburn and the Blessed Court are united in friendship and mutual protection."

"Yes, and the Court upholds the agreement to do nay harm." Mab's unnerving glance migrated from Beacon to Talfryn, then back with a steely glint. "But to my point, Talfryn would not be the first in your family to fall in love with a human."

Nettle stood on Talfryn's shoulder and shouted, "Talfryn is not in love with Fenella. He just likes her!"

"Not another word, Nettle!" Talfryn glanced at Lily. She averted her gaze, but not afore he saw the hurt and shock in her eyes. He sank lower against the wall 'til his spiky yellow hair was almost level with the thistle fairy.

"Well, Beacon?" Mab prompted.

Beacon narrowed his periwinkle eyes. His muted dark blue wings, spotted deep violet and rimmed first in amethyst, then in a narrow band of white, silently closed, then slowly parted. His voice assumed a bitter edge. "Are you referring to my beloved wife, dear Queen?"

Mab arched a sparkling tawny brow. The goblin stench of wet wool and damp dog collided with her frankincense, bitter orange, and cinnamon scent. Both intermingled with smoke and the flower fairies' sweet perfume.

Beacon's expression grew feral, his tone mightily restrained. "How could my pregnant wife, brought to Fairy against her will by you, dear Queen, as she began birthing, possibly have stolen the potion?"

"She left Fairy abruptly, did she not?" Mab's citric spice scent heightened. "As soon as your *Halfling* was born."

Beacon stilled like a cat poised to pounce. A sharp zing, the aroma afore a storm, cut through the cloying air. Blue lightning emanated from his fingertips, creating acorn-sized craters in the ground beside his feet. Those nearest him gave ground.

Talfryn's insides tumbled.

Chapter Four
Mab's Decree

Talfryn stood frozen as thin blue streaks emanating from Beacon's fingertips grew into violet-blue bolts. Clumps of earth flew, scattering fairies farther afield. He was on the verge of intervening when, with seeming sudden restraint, his father managed to quell his magic.

Beacon stressed his next words with a menacing stride towards the queen. "You insinuate my wife left of her own free will."

Mab huffed.

"All of Fairy knows you banished her!"

Silence descended upon the cavern.

"You tread on dangerous ground, Beacon." Mab's tone remained imperious, yet she seemed unnerved by Beacon's temerity as she turned to her consort. "Brindleburr, when was the last time I used the Aeternitas Potion?"

Brindleburr's eyes bulged. "I'm not sure," he stammered, then ran a shaky hand over his jet, slicked-back hair. "The marking of time is beneath us." Seeming to sense his answer insufficient, he straightened a bit and offered, "We could ask Glimmer if the potion has been used since the shield-maiden's banishment. Your sister dotes on details and trifling matters. It was she who knew when Beacon's wife was giving birth and suggested you retrieve the couple. And Glimmer remembered it was almost Midsummer."

Glimmer? My godmother betrayed my parents? Talfryn exchange a shocked glance with his father.

"Just ask the Royal Sister when Mab last paid tribute to the Unblessed Court," Nettle mumbled as murmurs rippled through the crowd.

Talfryn clapped his finger over the pixie's mouth. No one dared mention the tribute. And Mab had never explained why she sometimes paid a levy to a Court filled with kelpies, hags, bogeymen, redcaps, and malevolent elves.

The fairy queen raised one bronze, long-fingered hand for silence. "We need not involve my sister. I am certain the potion has not been used since..." The queen's disquieting gaze returned to Beacon. "What was your wife's name?"

Beacon's expression hardened. "Her name *is* Disa."

Sweat trickled between Talfryn's shoulder blades.

"Aye, Disa. The Norsewoman." Mab crinkled her nose. "She stank of Cold Iron even without her horrid weapons."

Talfryn tensed at the insult. Hot magic crackled down his own arms, twitching his fingers.

"While Disa was in Fairy, she was in the throes of childbirth," Beacon said through gritted teeth. "Onlookers drawn by her screams surrounded my home. They'll confirm they saw no one enter or leave Belle Flor 'til after you banished my son's mother."

Murmurs rose through the crowd. Many fairies inclined their heads in agreement. Stone-faced, the dwarfs remained silent.

The minute pale blue dot in each of Mab's eyes rolled into view. "Mayhap you stole the Aeternitas Potion and gave your *wife* the vial so she might become immortal like you."

A small bolt of enraged magic screeched from Talfryn's thumb. The silver-winged fay standing in front of him glanced over his shoulder and frowned. Mayhap fearing confrontation, the tiny thistle fairy flew off her parasol mushroom, then darted into the throng.

"I stole nothing." Beacon took another threatening step. Brindleburr blocked his way.

"Disa must have spirited the potion to one of those wretched Viking villages," Mab continued, ignoring Beacon's protests.

"Is that where you sent her, dear Queen?" Beacon's wings fluttered, stirring the tense air.

"I banished her to her own kind. Be grateful."

"Then none of us can retrieve the potion," Brindleburr said. "Cold Iron is poisonous to fairies."

"The dwarfs can recover it for ye." Vendel the dwarf stepped forward. *Indigite* dust from the crystal mines clung to his sandy hair and short scruffy beard. "Cold Iron doesn't harm us."

Why does the dwarf offer to retrieve the potion? Does he not believe my father when he says my mother is not the thief?

"Nay dwarf could enter a Viking village unnoticed," Beacon argued.

Father, too, concedes the potion is in the Norse lands. Is there truth in Mab's claim?

"Then I will send someone who can cast a glamour on the dwarf. Mayhap someone from the Noble House of Ruse." Mab's head swiveled, her pinpoint pupils sweeping those assembled 'til they focused on the fairies in front, who swiftly bowed their heads and averted their eyes. "Or better yet…" the queen glared at Talfryn. "You!" She stood and pointed her long finger at him. "Find your mother!" she spat. "And return my potion!"

Talfryn flinched.

Lily gasped and raised her hand to her mouth.

"My son is half fairy," Beacon protested. "Cold Iron can kill him."

Mab spun towards the Head of the Noble House of Storms. Her fists clenched. "His mother should have thought of that afore she stole from me!"

"If Disa had *somehow* possessed the potion and wished to become immortal, she would have drunk it while she was here," Beacon reasoned. "Then the potion could have been returned undetected."

"You credit your wife with more intelligence than humans possess." Mab's tone warned against further argument.

Inside Talfryn's pocket, his other pixie, Puffball, pushed back Talfryn's cloak with her spindly hand. Her pale green face poked out, and she blinked at the dozens of eyes fixed on Talfryn, then looked at Mab. "Dear Queen, mayhap Talfryn could engage the Thaness of Thorburn's aid. Lady Fenella is a trusted ally and oath bound to help us. She is immune to Cold Iron."

"Only to its poison," Talfryn objected. "She can be wounded or killed by Cold Iron weapons."

Mab shrugged. "I am willing to risk the girl's life."

"Humans are unreliable allies," Brindleburr warned. "They are ruled by their hearts."

"Then for all our sakes," Mab's icy stare drifted from Brindleburr to Talfryn, "you better hope Lady Fenella's heart belongs to you. Come dawn, I will magic you to her."

Dawn! Sunrise will be upon us within the hour.

Several smug gazes pierced Talfryn like elf shot. But Lily gaped at him with brimming eyes, her slender shoulders hunched, and her scorched wings drooped. He took an involuntary step towards her.

Puffball and Nettle mewed pleadingly.

Mab's regal nose twitched. "Very well, your sniveling pixies may accompany you."

Vendel stepped forward. "And I, dear Queen?"

Mab jutted her chin towards the sunken-eyed dwarf. "Go with them, and see they remain on task."

"Aye, dear Queen. As you wish."

"Beacon, if the Aeternitas Potion is returned by the full moon, then I *may* forgive your offense." Mab raised her chin. The amethysts on her dragonfly crown glinted. "If not—"

Talfryn stepped forward, "Dear Queen—"

"Meet me at the Stonefolk Circle at dawn!" Mab boomed, then vanished.

****

Dew had gathered like tears on golden buttercups and purple woundwort flanking the path to the Stonefolk Circle. At the top of the rise, Fairy stretched out around them. Forests claimed the north. Headlands met the sea to the east and south, and rolling hills dotted with fairy circles in the west were visible through a thin mist that grayed land and sky.

Talfryn waded through the vapors alongside his father and Vendel. He glanced about, hoping to glimpse Lily. Seeing no sign of her, a hollow ache bloomed in his chest. He scanned the trees for Luminous, although he knew the winged horse avoided Fairy and The Six. Spotting nay shimmer of cloud-white wings, nor hearing a welcoming whinny, Talfryn glanced at the pixies. They rode on Beacon's shoulder, clinging to his hair, as if they expected never to see him again.

A tiny part of Talfryn wished he could escape; run, then leap. Leap as far away as possible. *Did my mother really steal from Mab?*

How could that end in anything short of death or enslavement? How will I find her? And if I do, how will I survive her world of Cold Iron? How much time has passed in Fenella's world? Will she even remember me? But more important, how will I know when the full moon has arrived in the Blessed Court?

Beacon eased his stride and whispered in Talfryn's ear, "You must have many questions, but they cannot be answered here. When the time is right, use your magical gifts to discover the truths you seek."

Magical gifts? Do you mean my restored magic…or something new? He patted the charmed Finding Horn tied to his waist belt. His brow crinkled. *How will I know when the time is right? The full moon is in five days.* For all he knew, Mab thought it was in three days. Nerves had robbed him of feeling in his hands and feet. He glanced about the misty glen. *Nay sign of Lily.* His cheeks heated as he remembered Mab's warning that Fenella's heart had better belong to him.

Nay wonder Lily has failed to appear. Still, he had hoped. His throat thickened and hurt pressed against the back of his eyes.

Beacon stopped, turned towards his son, and removed the spun silver necklace he wore around his neck. Then placing it over Talfryn's head, he said, "Wear this for good health and protection." The crystal thunderstone charm, carved to a deadly curved point to resemble an equistragon talon, scratched a tear in his saffron shirt as it slipped beneath the fine wool fabric.

Father must be nervous; he placed the talon beneath my shirt, hidden from the world. Talfryn lifted the pulsing stone. "Nay Father, you must keep it. Only the Head of the Noble House of Storms may wear the family talisman."

Beacon stayed his son's hand. "Who knows what fate Mab has plotted? If the worst happens, I will rest easier knowing the thunderstone remains safe with you."

For the first time, Talfryn truly realized the weight of what he had known but had never allowed himself to imagine would come to pass; that should some tragic fate befall his father, he was next in line to lead the House of Storms. Then, should he fail to marry a fairy and produce an heir, the centuries-old Noble House of Storms would cease forever. He expelled a shivery breath, the thunderstone a cold, weighty knot of worry and responsibility above his heart. "I will return it to you."

Beacon clasped Talfryn's shoulders, then released him in what felt like a final farewell afore he strode on, picking up his pace.

Dazed and weak-kneed, Talfryn followed, maneuvering around blooming hawkweed and dew dappled lousewort. "Has anyone ever stolen from Mab and lived?"

"Now, son—"

"Mab rules by fear," Vendel warned in a low voice. "She cannot afford, nor is it in her nature, to be merciful."

"Surely, the noble families will protect you," Talfryn said.

"And risk Mab's wrath?" Puffball appeared close to tears.

Nettle patted her shoulder. "Don't worry. Remember, if a fairy kills a fairy, she loses her immortality. The queen is not immune."

"Mab would force someone else to do the deed," Puffball predicted.

"No one will be killed." Beacon stopped to face Talfryn.

"Someone killed Merrybelle," Talfryn countered and saw the brightness fade from Beacon's eyes.

"And if a fairy did the foul deed, we'll know soon enough." Vendel glanced up at Beacon. "As for Mab, aye, she would pressure someone else to do the killing for her unless she murdered in the heat of passion, without thinking."

"Mayhap, in a fit of jealousy?" Nettle asked.

"Aye. Or if someone provoked her, and she snapped." Vendel's flinty gaze held Beacon's.

Talfryn sensed an undercurrent—pointed and foreboding like a coded message—pass between his father and the dwarf.

"Have courage," Beacon told Talfryn. "You will succeed. Always remember, you are a Trooping Fairy. You are of noble blood."

"Half noble blood."

"You have my magic, your mother's strength and courage." Beacon's tone was fervent, yet fatherly. As he resumed the march with a more urgent, determined stride, he avowed, "Nothing can stop you."

Except Cold Iron. A shiver rattled through him as he remembered the high fever, tremors, and loss of awareness he had suffered when he had contracted Cold Iron poisoning but one year ago at Beltane. *Had my mother's Norse blood not flowed within me, and had Mab's sister, Glimmer, not whisked me back to Fairy, I would have died.*

"Search for Disa in the northern Kingdom of the Isles. The dwarf who raised her is from one of those isles, and your mother would have returned to him if possible. She had journeyed farther north when I first encountered her on Skye. I'm not sure which century."

Abruptly, Beacon halted again, and they all fell in line next to him.

They had reached the small standing stone circle. Talfryn scanned the ancient rocks. One had a deep notch. Lichen grew in a spiral pattern on another. Blooming lousewort stood out against the gray stones.

Beacon removed a narrow ring of twisted gold from his finger. "Give her this. Be sure to wear her brooch on your cloak.

And bare your wrist cuff so she'll know I sent you."

Talfryn nodded. Half reluctant to take his father's wedding ring, yet half hopeful at the prospect of meeting his mother for the first time, Talfryn slipped the band onto his finger.

"She'll recognize your eyes," Beacon assured him. "They're the same gray as hers. Of course, Disa's do not tilt up. Her ring bears a blue tourmaline set in gold and two smaller stones." Beacon frowned. "Are you certain you wish to wear that saffron shirt and so little else? You will freeze."

Talfryn stood taller though he felt uncomfortable wearing such bright colors on such a terrible day. Vendel, he had noticed, wore a padded, pale fawn overtunic and leggings. "I need to blend in with the Highlanders when we reach Fenella."

"Ah, aye." His father's wings swept together, then parted. "Good thinking."

Beacon sighed. "At least you have your magic cloak. And the Finding Horn?"

"Aye." Talfryn opened his cloak and revealed the smooth surface of the enchanted vessel. "But Father, what if I fail? Mab will—"

"You will not fail. And when you return, you'll resume your magical studies." Beacon pressed a small, blue-violet silken pouch into Talfryn's hand. "Use its contents—"

Strong spicy scents choked the air.

Talfryn tensed, then slipped the weighty little pouch into his cloak's hidden pocket.

In a hot whirl of gold dust, Mab appeared beside them. She brushed glittering specks from the ebony silk webbing covering one of her bare shoulders. The specks gleamed red-gold against the ash-gray rock at her feet, glowed bright, then glinted out. "All present?"

"Aye," Beacon said.

The pixies leapt from Beacon's shoulder and scrambled into Talfryn's cloak pocket.

"Good." Mab swept backwards and flicked her hand.

A tree with fog-gray bark and sun-hued leaves materialized outside the stone circle. For a moment, it swayed. Shock, pain, and bewilderment seemed to emanate from its core. Earthy scents filled the air as the rowan threw down deep roots to steady itself.

Mab raised her hand again, retreating a step as though she expected something dangerous to come forth.

Instinctively, Talfryn, Beacon, and Vendel shrank backwards, as well. The pixies scrunched down until only their eyes, and their small hands gripping the pocket edge, were visible.

When Mab flexed her fingers, a grim-faced dwarf appeared. His grubby tunic, the same rose brown as blusher mushrooms, hung loosely on his frame, as if he had grown too thin for the garment. Dirt smudged his high forehead and clumped his gray hair. He cowered afore Mab.

"Bleak! Do as you were told," the queen snapped.

The dwarf trembled. With an apologetic glance towards Beacon, the miserable creature pulled a heavy rope from his sack, then looped it around the disoriented tree.

Engrossed with the dwarf, Talfryn almost missed the next flick of Mab's hand. A glowing metal cuff encircled his father's right wrist, and with a loud click, locked.

Using uncanny speed, the dwarf cinched the rope to the cuff with a Cold Iron link.

Beacon hissed in air between his gritted teeth. "This isn't necessary, dear Queen."

Mab extended her palm, halting his words. "You are hereby Fairy bound without magic," she decreed, "'til your son returns with the potion."

Talfryn clasped his hands together to force down the angry magic sparking beneath his fingertips.

The spirit within the tree moaned as the dwarf withdrew more Cold Iron links from his sack.

"Every two days, another iron link will be added," Mab warned, "'til the potion is returned…" She raised a glittering brow at Talfryn. "Or your father succumbs."

Fury and fear propelled Talfryn forward, but his father shook his head. Fists crackling with magic, Talfryn grudgingly held back.

"Everyone ready?" Mab asked, as though they were all gathered for a pleasant picnic.

Vendel patted his tunic near his hip. "Aye."

Talfryn tore his gaze from his father, then narrowed his eyes at Mab. "Aye."

"Good. Pray your Lady Fenella yet lives. Return afore the full moon rises in the night sky!" Mab sketched a symbol in the air.

"Wait! How many days will that be in human time?" Talfryn asked, panicked.

Afore Beacon could answer, a blistering wind pushed Talfryn, Vendel, and the pixies between the portal stones. As Talfryn stumbled into the Stonefolk Circle, his father, the Queen of Fairy, and the dwarf clutching Cold Iron links vanished in a blinding burst of hot light.

CHAPTER FIVE
FENELLA

June 1506
The Blackwood Stronghold, Scottish Lowlands

A flash of bright morning light broke through clouded skies and beamed its warm ray on Lady Fenella, the Thaness of Thorburn, as she pushed open the woven willow gate and slipped into the medicinal garden hidden behind her captor's towering keep. Aware of her two ever-present shadows, Calidore, her massive Blackwood guard, and Derry, her peculiar maid and unlikely chaperone, Fenella inhaled the loamy scent of tilled earth and soothing fragrances of rosemary, dill, thyme, and lavender. The scents transported her back to her beloved Highlands, her home, and her clan, all of which Lord Argonshire had stolen from her the day his men had seized her.

Fenella glanced at the tansy and other poisonous plants with curative properties that grew in a square plot in the far corner. For a heart-skipping second, she flashed on Fairy and the maze garden at Belle Flor.

Unconsciously, she put her hand to her throat, and the Thor's hammer charm dangling from a gold necklace beneath her léine slid across her heart. *Talfryn is dead*, she reminded herself. *He must be, or he would have revealed himself months ago.* Grief and disappointment warred within her as she tore a faded flower from a lavender stalk and crushed it between her fingers. Fenella watched the pieces fall like fairy dust onto the turf path.

Drifting deeper into the garden, she came upon the fountain where spring water spilled from a stone lion's mouth into two small, successively lower pools. A warbler, perched on the ledge, dipped its beak into the water. Watching the small bird, Fenella's thoughts wandered back to a year ago, after her escape from Fairy and her return to the Blackwood fortress…and Edward.

In the shadow of the donjon, she had collided with Derry.

"Many pardons, my lady!" The maid swayed sideways and dropped her basket of freshly picked mint and thyme. Calidore clasped Derry's slender arm and steadied her. The young woman beamed up at him. "Thank you ever so much." Then evidently remembering her mistress's presence, Derry lowered her voice to a loud whisper and said, "Mr. Blackwood requests your presence in the…hidden garden."

Confused, Fenella glanced at the kitchen garden to her left.

"Nae that one." Calidore turned to Derry and said, "Deliver your basket to Cook. I will show Lady Fenella the way."

"These herbs are nae for Cook. Mr. Blackwood ordered them sent up to Lady Fenella's chamber."

"For me?" Fenella asked.

"Aye, my lady." The maid waved a flat-leafed sprig under Fenella's nose. "The mint has restorative qualities, and the thyme is to be sewn into a pillow to ease your nightmares."

"I see." Fenella blushed. Had she cried out in her sleep again? Had the night guard outside her door informed Lord Argonshire, or just Mr. Blackwood? Countering Calidore, she said, "Derry, we shall wait 'til you return."

The maid bobbed in her unsteady way, sending her dull, seal-brown tresses cascading towards her narrow chest. She lifted the basket, then squared her shoulders afore disappearing into the stronghold.

Fenella shook her head.

Derry returned so quickly, Fenella wondered if she had sprouted wings and flown up to the tower chamber.

"Ready," the maid said, breathless.

"This way, my lady." Calidore strode past the fenced vegetable and herb garden. Young grooms releasing steeds to graze in the courtyard cast curious glances at Fenella afore nodding to Calidore. None appeared to notice Derry, who sidestepped to avoid the horses.

Fenella jutted her chin and quickened her stride. Following Calidore around the northwest tower, her skin prickled as she plunged into a cool grove of thick-trunked yews. She scanned the woods. As much as she trusted Calidore and Derry…

"Who told you Mr. Blackwood requested my presence?"

"Why, Mr. Blackwood," Derry answered, her expression puzzled and without guile.

"I have never been back here afore." Fenella kept her voice neutral and slowed her pace.

"These days, 'tis rare for the Blackwoods or their…guests to venture behind the stronghold," Calidore said.

"Guests are free to leave whenever they wish," Fenella retorted, eyeing her two escorts in defiance. "I am nae a guest."

Calidore and Derry exchanged a discomfited glance, then Derry lowered her gaze.

The seclusion and silence pressed against Fenella as she surveyed the strange surroundings. Tree canopy blocked whatever windows might exist on the back of the donjon. A blackbird flew from one yew to another. Gran had said blackbirds foretold magic and knowledge.

Fenella shivered. Out of the corner of her eye, she thought she saw a wolfhound slink between the trees.

She emerged from the grove into a sunny clearing where shimmering blue butterflies flitted. "How beautiful!" Reminded of Fairy, Fenella strode ahead for a closer look. Seeing the creatures were indeed winged bugs and nothing more, she released a disappointed sigh.

Calidore crossed the clearing to a fenced enclosure, opened the woven willow gate, and peered inside. Apparently deciding all was safe, he waved her in.

Derry silently clapped her fingertips together.

Fenella paused just inside the gate. The heady scent of roses welcomed her. Bees buzzed around a pair of rose bushes, and the sound of flowing water reached her ears. She gazed down the turf pathways separating raised square plots. In the closest flowerbed, scraggly lavender and leggy rosemary grew alongside hyssop that had yet to bloom. Magic eddied beneath her feet, tingling her skin.

"Mr. Blackwood?" she called out.

"Here, Lady Fenella."

She glimpsed Edward deep in the garden. When she had been bedridden with fever and shock after his father's men had stolen her, Edward's strapping appearance had jolted her. But his kindheartedness softened his features, and his broad shoulders and chest now elicited comfort, nae alarm.

Edward turned towards her for a mere second afore returning his gaze to the fountain. Her heart squeezed at the slight. Unsure of his intent, she tracked her thumb across the key-shaped scar on her palm afore setting a determined stride down the spongy turf path.

"Mr. Blackwood," she repeated when she reached him.

This time he pivoted, his injured leg bending awkwardly. He winced and tightened his grip on his walking stick. A lock of his raven hair swept one eye.

Remembering his injury, incurred because of her, her heart softened as she reached up to inch the errant strands behind his ear.

Edward's breath hitched. He covered her hand with his and pressed her palm against the rough stubble along his jaw. His gold-flecked eyes held her gaze.

Warmth spread like a sunrise in her chest. "You wished to see me?"

"Aye." As though realizing his gesture too familiar, he released her hand. "Afore my brother left—"

"Hamish the monk or the wayward golden son, Gavin?"

His expression eased. "Hamish. Gavin never gave a thought to anyone else's wellbeing. Hamish studied plant lore. Herbology. He loved this garden. He worked the soil. Pruned. Harvested. Sometimes, he just sat on that bench." Edward angled his head towards the seat behind them. "And stared at the fountain." He stooped over his walking stick, favoring the leg with an infected dog bite.

"Did you join him?" Fenella asked.

"Nae often. I sensed he desired solitude." A regretful-sounding sigh fled his mouth. "But sometimes, he would tell me tales aboot the fountain."

"What sort of tales?"

"Daft ones. He half convinced me the stone lion could speak, and if I were verra still and patient, a lady might appear in the water."

The back of Fenella's head tingled as she slid her gaze to the lion spouting water.

Blackwood shook his head. "You will nae see nor hear anything. I never did." He released a long sigh. "The garden seems…restless now. Lonely. Different somehow." His eyes searched hers. "Do you think me mad?"

Magic pushed up from the ground like fast growing vines and twitched Fenella's toes. She stomped her feet. The enchantments stilled. "There is an air of neglect, an absence of love. Mayhap that is what you sense. Who tends the plots now?"

He shrugged. "I am nae certain."

Fenella waited. Unsure of what he wanted. Needed.

Edward took her hand in his, "Take charge of the garden." His soft-spoken words seemed more a plea than a demand.

She gaped. Did he wish to further reduce her tenuous standing? "Surely you are nae serious?" Fenella had considered him, her captor's son—her betrothed—to be her sole friend at the stronghold save for Derry, and mayhap Calidore in his solemn way.

Enchantments tugged her hem. Was this a test to reveal her taint? She slipped her hand from his.

Blackwood cleared his throat. "Nightmares have again stolen your sleep."

Fenella flinched at the conversation's unexpected turn. "The night watch outside my chamber has keen ears and a loose tongue." Off guard, Fenella hugged herself. "I doona sleep because I doona wish to dream." She subdued the goose flesh rising on her upper arms. "In my sleep, I relive the day your father stole me."

Blackwood sighed. "I suspected as much. Which is why I ask you to tend the garden."

"You confound me, Mr. Blackwood!" Anger, at the affront, heated her cheeks. "To what end do you ask this of me?"

"Hamish believes working with plants soothes the mind and spirit. 'Tis what he would recommend if he were here." Blackwood sat heavily on the small wooden bench and gestured for her to sit beside him. When she had settled, he clasped her hand once more and lowered his voice to an intimate whisper. "'Twill spur your healing, Lady Fenella. Our healing." His warmth soaked into her skin.

"Verra well," she relented. Alone for so long, she felt a sudden desire to be close with someone…to find solace…with him. He had

favored her afore her…misadventure. "You may sit on this bench and direct me."

Blackwood chuckled. "You were born to give orders, Thaness, nae receive them."

"Then this shall be a good lesson for me." She squeezed his hand. "Together, we will vanquish the darkness clouding our souls."

He regarded her for a moment. His jaw twitched, as though he were weighing feelings of affection he had once felt for her against pangs of betrayal caused by her recent mysterious disappearance.

"Hamish would like you," he said at last.

"And I, him."

A distant rustle caused the warbler to take flight, shaking Fenella from her reverie. She twisted towards the low wall encircling the garden. Hawthorn and crabapple saplings, she and Edward had planted along the barrier, would someday overtake the wall and afford them some privacy. But for now, she had an unobstructed view.

Edward emerged from the yew grove. His gaze sought hers. A thrill surged her like a breaking tide. If her eyesight had been less keen, she would have missed the slight, secret upturn of his lips when he spotted her.

Even here, seemingly beyond your father's prying eyes, we must pretend there is little fondness between us. For I have nay doubt only a loveless marriage will serve his foul purpose. Glaring up at his chambers beyond the trees, she silently directed her ire at Lord Argonshire. *You need me to marry one of your sons, doona you? A son you can control. A son who will nae rise against you should you murder me after the wedding. 'Tis my luck, you have greatly underestimated Edward.* Though she could never address him thus aloud, it felt freeing to vent her anger, reveal she knew full well his evil intent, and thwart him with Edward's love for her.

His lordship's spies infested the stronghold. She glanced higher at the ramparts. A guard, arms crossed against his burly chest, stared down at her, as though he had heard her every thought.

Fenella dropped her gaze. Feigning nay interest in the youngest Blackwood son, she whirled away from him. With casual strides, she followed the turf path to the strawberry plot. After picking several of the plump, heart-shaped berries, she followed a bee as it buzzed towards the roses flanking the garden gate.

Calidore and Derry retreated a discreet distance as her betrothed strode towards them with his hand pressed to his thigh. His limp appeared much less noticeable than it had a year ago when his father had refused him treatment for the dog bite. Anger flared anew when Fenella thought of how Argonshire had punished Edward for her escape.

How could a man as heartless as Argonshire sire such a commendable son?

Edward drew nearer as if pulled by their furtive bond.

Though all the Blackwoods save Hamish were raven-haired, so Derry had told her, Fenella suspected only Edward had eyes that changed with his mood from deep sage to golden brown flecked with green or gold. *Eyes like a summer garden.*

"How grows the fruit of your labor, Thaness?" Edward asked.

Fenella offered up the strawberries. "It shows great promise, Mr. Blackwood."

He lifted his gaze from the lush fruit to meet her eyes. "And great beauty."

She felt a warm blush fan across her cheeks when she saw her own reflection pooled in the bright green and gold flecks of his regard.

Without breaking their locked gaze, Edward reached for the berries. Their fingers touched, fluttering her core. Her world narrowed down to their hands. Edward brushed his thumb across her flesh, leaving a tingling trail.

After a moment, he cleared his throat and shifted his attention to the garden. "You have worked wonders."

"*We* have worked wonders." Fenella glanced back with pride at the blue-violet hyssop and borage blooming next to cabbage and heartsease. Bright marigolds edged beds of parsley and mint. Carrots waited harvesting. "I am glad you convinced me to take charge."

A rare smile lifted his lips. "We vanquished the darkness."

At the verra least, we lifted the darkness between us. She returned his smile. "Aye."

"Hamish would be pleased. You two share the same magical touch with plants."

At the word, "magical," Fenella inwardly winced. "I am indebted to Hamish for teaching you herbology and the healing arts. He must have joined a remote monastery to have never returned home."

Edward frowned. "Aye. He must have."

"You doona know where he is posted?"

He shook his head.

How peculiar. Spotting one of Argonshire's men striding towards them, Fenella frowned and stepped back. Edward glanced over his shoulder, barely turning his head. His broad shoulders sagged, and Fenella noted the scowl he quickly masked.

"My thanks to you, Lady Fenella," he said in a formal tone as he held the strawberries aloft. "Farewell."

The man was almost upon them. He wore the familiar *your-father-requires-your-presence* expression.

Fenella downcast her gaze. "Farewell, Mr. Blackwood." As Edward twisted away, she wondered what excuse his lordship had conjured this time to keep them apart.

Vexed and brooding, Fenella joined Derry by the fountain, passing Calidore as he resumed his post at the gate. The garden's magical spell and its illusion of freedom had vanished.

Chapter Six
The Way In

Three mountains loomed afore Disa, and as she matched the shapes against the seer's map, a dangerous spark of hope ignited. She willed her speeding heart to slow. After traveling weeks by boat, horse cart, then foot, through ground clouds and sleet, after following false leads, fighting off knaves and scoundrels, she dared not believe she had finally arrived. Every moment lost had been another moment Disa fretted for Bree's safety. Her heart could not survive more pain.

Once again, she meticulously compared her surroundings to the map. "Thanks be to the gods and goddesses." She dropped to her knees. "I found it, Bree," she whispered to the air as she wiped her eyes with the hem of her woolen cloak. "I found the island."

Disa struggled to her feet, suddenly bone tired, and tucked the sheepskin inside her quilted tunic. Afore her, the westernmost mountain appeared exactly as Ragna had drawn it—steep-sided and barren save for the trees skirting its base. An equally close, but lower, more approachable mountain loomed in the east. Yet neither peak called to her.

But between the two mounds, in the far distance, a third mountain beckoned like a whispered secret. And an invisible force, as inescapable as a fisherman's long line and hook, drew her to a narrow pilgrim's path. She adjusted the circular shield slung over her shoulder and strode towards her fate.

The path began as a gentle climb. On the moor below and to her right, a river twisted in and out of view. Sometimes wide and thunderous, it stormed against steadfast rocks. Other times, it disappeared almost entirely, trickling quietly, half hidden in bog grass.

When water from unseen springs flooded the rocky path, Disa slowed her steps to heed her footing. She passed through lush glens and rolling hillocks where more than once she paused, glimpsing a fairy ring on an emerald slope. A lilting fay song wormed into her ears. Simple rhymes strung together in musical bits and pieces. Disa began to skip, her feet nay longer cold nor tired. She glanced longingly at each fairy ring. She could almost imagine Bree dancing among the stones with her fairy cat, Muirgen. But the mountain's magic drew her onward.

Each time she rounded a curve, though the landscape ahead changed, the distant peak never grew closer.

Gray clouds scuttled across the sky, momentarily shrouding the far tor and blocking the sun. Disa shivered and drew her wool cape more closely around her. Moments passed, then the clouds parted. She blinked at the distant peak glistening white as though suddenly blanketed in snow.

Lost in a fairy song, Disa barely noticed the hours passing or the mist creeping between the foothills. More dove-gray clouds settled in the sky. Reaching for her water flask, her hand brushed against the iron hilt of her sword. The fairy voices singing inside her head halted.

Alarm rose like a sudden flame. *Where am I?* Disa glanced about. *I should not be on the path after sundown!* Footsteps sounded behind her. She whirled. Seeing no one, her skin prickled. Weary, yet watchful, she had no choice but to press on.

The trail rounded a series of shimmering pools. Disa bent to fill her flask. In the clear water, her image stared back at her—wary gray eyes, fierce mouth, blond braid clawed by the wind. "You are a sight," she muttered.

Her words rippled through the glens. *You are a sight. You are a sight. You are a sight.*

She stood so quickly she saw stars. Pulling her cape close, Disa scanned her surroundings. *Nay fay folk along the path. Nay snatchers in the sky. Yet, I sense a presence.* As she continued on and approached the next pool, the back of her neck tingled. Barely breathing, she eased her hand towards her sword. Without turning her head, she eyed the sparkling water, spying the reflection of a large brown bear close at her heels. With a gasp, Disa unsheathed her sword and twisted to confront the beast.

But the bear had vanished.

Disa gaped. Hlín's words came back to her. *"Your soul's protector, your fylgja, will materialize to attend you at the time of your death."*

"Nay," she barely breathed the word. Then with a force of strength drawn from deep in her core, she commanded, "Stay away, *fylgja.* Today is not my death day."

Again, her words echoed through the passes. *Today is not my death day. Today is not my death day. Today is not my death day.*

The path appeared empty. But so persevering was the bear's attendance, Disa could almost feel its breath on her nape. Her fingers twitched on the sword hilt as she scanned her surroundings.

"I thank you for protecting my soul," Disa said in a softer tone. Then, to prove there was nay danger, she sheathed her sword and resumed walking. But still, she sensed the spirit bear padding behind her as, once more, fairy music crept into her ears.

After what seemed hours, her steps deadened with fatigue, Disa reached the foot of the elusive mountain. The fairy tunes had grown louder, faster. She clapped her hands over her ears in a fruitless attempt to silence the fay voices.

Behind her, the spirit bear rose and pressed its blunt, curved paws over her hands. Disa froze as sudden silence replaced the fairy song.

When nothing more happened, fear mixed with perplexity. *Is the bear preparing me for death, or protecting me from it? Hlín's foreboding message made nay mention of such actions. Is this enigma what one feels afore one dies?*

Disa's thoughts sharpened. *The bear has one purpose. Why is it still with me? And why has nay mortal threat appeared?* An urgency of mission grappled with this uncertainty. *I have nay control over such matters as Fate.* She forced her mind to clear itself of dismaying thoughts. *These feelings of impending doom only hinder my task. I need to find my way into Fairy; track down Queen Mab.*

Suddenly, high on the side of the mountain, a portal opened with a blinding orange flash. Disa blinked, not trusting her eyes, as through the breach, a winged, blue-skinned figure catapulted downward, hurtling directly towards her. His arms flailed and his wings, not angel wings like the statue in front of the monastery of her early childhood with Bree, but glorious, expansive butterfly wings, flapped in a blue-violet blur, beating the air with what seemed an impressive and powerful effort to gain control of his plummet.

With her heart pulsing furiously, Disa pushed the bear's paws from her ears. Casting aside all thought of the fylgja, she trained her gaze on the fast-descending blue man. Lowering her hand to her sword hilt, Disa stood her ground, too entranced to raise her shield.

Spotting her, the being flinched. Through some trick of the light or fairy magic, his wings shriveled and vanished; his skin became as fair as hers. He reared midair like a horse roughly reined in. With legs stiffened, he flexed his toes skyward as though trying to halt midflight. Then he dropped from the sky, landing roughly on his feet, and with great forward speed, stumbled several steps afore barreling into Disa. Scooping her up in his arms, he held her secure in his tight embrace as he skillfully skidded to a stop.

His bright azurean gaze locked with hers. Sudden warmth flushed her skin as his bracing breath washed over her, fragrant as lavender and mint mixed with hints of rosemary. Inexplicable relaxation stole through her body as she wilted in his arms; the aches and pains of her months-long journey melted away. Transfixed, Disa hardly noticed the swirl in her stomach as her *fylgja* retreated.

Afore Disa could close her gaping jaw, the enchantment between them severed when a beast with shimmering scales, two tall, twisted horns atop its head, and enormous bat wings swooped to the ground behind the stranger. It flexed its front talons afore landing on its two hind horse legs. Small whirlwinds eddied around its hooves. Keen intelligence gleamed in the strange beast's eyes; it regarded her with an unnerving stare.

In one fluid motion, Disa broke free from the stranger's strong grasp, whipped her shield around, and drew her sword.

The man frowned, retreating several steps.

Behind him, the portal on the side of the mountain swirled closed, swallowing the bright orange light. The creature puffed smoke out its nostrils, then lowered its jaw, revealing deadly dragon teeth.

With great effort, Disa downed the hard knot of fear lodged

in her throat. Her attention jumped from the mountain, to the dangerous beast, then back to the man. "You are not Norse," she said, taking in his slender build and lack of weapons, markings, or battle scars. And not forgetting for one moment, though no longer visible, those magnificent wings.

"Nay, but your fair coloring and—" His gaze lingered on her eyes, then dropped to the Thor's hammer charm she wore around her neck, afore traveling lower. "—well, your weaponry, suggest *you* are."

"I am half Norse," she corrected, somewhat bewitched by the unexpected graciousness in his deep voice yet unnerved by his obvious appraisal of her. "My family was lost to me when I was young. I was raised by…another." She blinked back unexpected tears.

The stranger regarded her. Not with a predatory gaze like the men she had previously encountered, but in a thoughtful way, as though he were working through the implications of her statement. When his tilted eyes widened, Disa thought, *I revealed too much. What truth has he surmised?*

Her heart thundered. *I've been sent to a land where men's beauty rivals the gods' and horses resemble dragons.* She shifted her weight from one foot to the other. *Did I choose the wrong mountain? Follow the wrong fate?*

"We appear destined to meet," the stranger said as he set about brushing the sleeves of his sky-blue tunic, then straightened it over his tan leggings.

Destined? She lowered her shield.

Satisfied with having put his clothes to rights and with a slight bow of his head, he presented himself to her. "I am called Beacon." His tilted blue-violet eyes, keen and curious, glittered.

"Why do you believe we were *destined* to meet?" she asked, not revealing her name.

"My equistragon—" He jutted his thumb towards the fearsome horse-dragon grazing behind him. "—has never thrown me afore. And if he hadn't, I would not have discovered the wormhole in the side of the mountain, nor encountered you." He pushed his snow-white hair off his youthful face; apparently unaware or unconcerned of exposing his elegant yet striking pointed ears.

Fairy ears. "Wormhole?"

He swatted the air in a dismissive gesture. "A magical portal."

So Ragna's map proves true. The confirmation swept through her. Remembering the fairy rings she had passed and the mountain wormhole she had just seen, Disa thought, *Secret entrances to the fairy realm must be all around me. My finding a man such as this surely has reason.*

Beacon's eyes gleamed, reminding her of Muirgen, the fairy cat. *Secrets and mischief. Here is a man who knows things, a man who can lead me to triumph.*

With an elegant gesture, he extended his hand. "And you are?"

"Pleased our paths crossed." Disa sheathed her sword and swung her shield onto her back. She stepped closer and clasped his forearm. An unexpected, though not unpleasant, tingle shivered her core.

She angled her head towards the mountain. "Is that the entrance to Fairy?"

He studied her a moment. "It is one."

Disa tried to avert her eyes to break the fairy's captivating gaze—but to little avail. *I will not be beguiled.*

Flustered, she released her hold on his arm and fingered the leather pouch tied at her waist. Hidden within the sack, the tiny crystal sphere that the dwarf's mage had given her pulsed reassuringly. She could feel the worn edges of Bree's heart-shaped yellowish-green stone and Hlín's birch rune.

Grambolyn's words repeated in her head. *"Find proof of Queen Mab's foul deed or find the means to kill her."*

Squaring her shoulders beneath the heavy shield, Disa stared into the fairy's entrancing eyes with newfound resolve. *Somehow, I must find my way to Mab. If meeting this man is truly destiny…*

Guarding against the magic fluttering her heart, she calculated her next move.

Chapter Seven
Messengers

Fenella sat at the high table, the only female, the only redhead, and the only captive in the teeming Blackwood hall. Like a hen among wolves, she was mindful of the Blackwood garrison. Seated at side trestle tables afore her, the men hunched over their trenchers, eating and drinking. They bantered and clapped each other on the shoulder. Of the threescore of them, who a year ago had stolen her from her stronghold and clan, nae a one had showed remorse. *Ever.*

She tightened her grip on her goblet. Though 'twas fruitless, she glanced at the two escape routes that led outdoors. Both the main staircase near the dais and the screens passage that separated the Great Hall from the service quarters at the opposite end of the hall led down to the upper courtyard.

Sipping her spiced wine, Fenella fought warring urges to remain vigilant or flee to her isolated chamber.

Doona appear weak. She straightened her back, conjured a composed demeanor. Bone tired and brittle from a sleepless night listening for Edward's contingent, Fenella imagined a litany of ill fates that might have befallen him. Worry vanquished her hunger. She stared unseeingly at the candied almonds and wedge of fruit cheese that comprised the final course of the late morning meal.

Fenella worried her thumb across her scar. *If anything has happened to Mr. Blackwood…* She narrowed her eyes and slid a sideways glance at Lord Argonshire, the only other occupant at the high table, save for a suspicious-eyed priest. Anger and dislike flushed her skin. *I will blame you.* For surely, Argonshire had sent his son off on another pretense—trouble at a Blackwood holding, or mayhap a sighting of his lordship's long absent eldest son, Gavin.

You will do anything to keep Edward and me from becoming close.

Although his lordship had stolen her to gain guardianship of her and reap the wealth of her thanedom, his control legally ended when she married or turned ten and eight. She was certain Lord Argonshire had nay intention of relinquishing those reins.

Her nape prickled when she felt Lord Argonshire's gaze land on her like a hawk snatching a rabbit, piercing her with obvious suspicion and disdain. As Fenella's discomfort intensified, she fought the urge to shift position and angle her body away from Argonshire. *Does his lordship watch for proof I carry the taint?* Fear coiled within her, but she would not give him the satisfaction of seeing her cower. She adjusted her mantle, wrapping the length of scarlet wool more tightly around her, but maintained her countenance.

Thankfully, two empty seats separated Fenella from his lordship and the clergyman. Nay chair awaited Argonshire's middle son, Hamish. But a chair did await Gavin, *He-Who-Can-Do-Nay-Wrong. Do Gavin and Hamish share their father's arrogant nose and distrustful brown eyes? 'Tis doubtful their eyes would reveal a lifetime of rejection, like Edward's.*

She surreptitiously reached beneath the table and smoothed her hand across Edward's empty chair, which was nearest her. *If further harm has come to you, Mr. Blackwood, I will avenge you.*

Somehow. Someday. Her freckled face heated.

To calm her nerves, Fenella reviewed her secret plan. *Once I have unshackled myself from Lord Argonshire, I shall make him suffer for stealing my freedom, for seizing my ancestral lands, and for his ill treatment of his third son.* Nae a soul kenned of her plan, nae even her cousin, nor especially Edward Blackwood. All her hopes hinged on marrying Blackwood for love. *Then, should I survive our wedding day, I will convince Blackwood to escape with me to my Highlands.*

While she fingered the stem of her goblet, her gaze flicked once more to the main staircase. Her chest tightened. *Where are you, Mr. Blackwood?*

Lifting a nut to her mouth, Fenella's thoughts shifted to her greatest threat. *A year has passed since Viviane cast a spell to prevent my scar from vanishing.* Fenella was nae sure how much longer the Lady of the Lake's enchantment would last. Without her distinct scar, she could nae prove to her clan she was the true Thaness of Thorburn.

"Stop frowning," Lord Argonshire commanded, his voice low and threatening. "The men are staring."

Fenella's jaw dropped, and her face blazed anew. After her year in captivity, or "safekeeping" as Lord Argonshire referred to her imprisonment, the garrison largely ignored her presence. But a quick scan of the hall confirmed his words. Most of the men lowered their gazes, but some, mayhap emboldened by Blackwood's absence, cast stares that crawled her skin. Calidore, seated at a nearby trestle table, visibly tensed.

She cleared her throat. "Has there been any word from Mr. Blackwood?"

Lord Argonshire lowered his ale cup a fraction. "Why?" He eyed her. "Have you grown fond of him?"

Her pulse surged with indignation. *Aye. More than you can imagine.*

A small commotion erupted near the entrance porch behind the screens passage at the far end of the hall. Fenella's breath caught in anticipation. Wind gusted through the hall, rustling the black banners slashed with crimson that hung from the rafters. Candles on the antler chandelier flickered as a member of the Blackwood first watch entered the large chamber. The man made straight for the dais, striding past his kinsmen seated at the trestle tables without a word of greeting. Strewing herbs crunched and flattened beneath his boots. When he neared the high table, the watchman halted.

"Speak," Argonshire said as though the guard were a well-trained hound.

"Two messengers request your ear, my lord, but they dinna arrive together."

Argonshire placed his ale cup on the table.

"One arrived from the north," the man continued, glancing at Fenella, "the other from the south."

The north? Could the messenger be from my clan?

"Admit the one from the south, and send him to the oriel chamber," Argonshire instructed as he pushed back his chair and rose. "Have the northerner wait outside."

"Aye, my lord." The Blackwood guard turned and retreated to the screens passage.

"Hasten to your chamber," Argonshire ordered under his breath, as though she were five instead of weeks away from turning ten and six and becoming a bride.

Fenella ground her teeth. Had Argonshire nae stolen her, she would be at *her* castle, where *she* gave the orders. Biting back rage, she stood and said, "As you wish."

As she left, Fenella heard the side door behind the high table click shut, indicating Argonshire had taken his leave. Calidore rose from his bench. Fenella, with a glare and slight nod, bid him remain. Frowning, Calidore waited with his powerful arms folded over his chest.

Instead of retreating to her chamber, Fenella headed to the screens passage. Argonshire's men stared as she strode past their tables. She silently cursed the blush warming her cheeks. Calidore fell into step behind her. Relieved the mountainous guard had ignored her command, she lowered her tense shoulders and led with her chin.

Fenella slipped past the carved oak screens and entered the passage. She heard muffled servants' chatter behind the closed door set into the back wall. Skirting it, she turned right, and hurried towards the entrance porch.

The watchman had left his post, presumably to retrieve the messenger from the south. Another member of the garrison was closing the door to the outside stairs.

"James!" Fenella called out.

"I ken what you are planning," Calidore whispered.

She ignored him and strode up to the young guard. Fenella had pointedly learned as much as she could about the watchmen. "Greetings, James. Did your wife receive the feverfew I sent?"

"Aye, thank you, my lady. Derry brought it to us yesterday. My Bridget's headache had much improved by morning."

"I am glad to hear of it." She glanced over his shoulder. "I understand a messenger has traveled here from the north—"

"Aye, the same one from ten days ago. He asked for you again."

Fenella's heart quickened. "I was never informed of a visitor. Where is he?"

"Thaness," Calidore warned, his voice overly loud as if he meant for it to carry outside.

"Lady Fenella?" A familiar male voice rose from the upper courtyard.

"Uncle Connell!"

Fenella slipped past James, ran down the outer timber stairs, and launched herself at her late father's brother. Connell engulfed her in an embrace so tight 'twas as though her whole clan hugged her. He smelled of horse, wool, and early blooming heather. A longing for home washed over her.

"Are they treating you well, lass?" Connell examined her with his keen green eyes. He had aged in the year since she had last seen him. Iron-gray strands threaded his thick brown hair. Afore she could answer, he asked, "Did you receive my last message?"

His tight expression plummeted her heart. "Nay! I have only now learned of your previous visit. What news do you bring? Is everyone well?"

Connell opened his mouth to answer, then paused when Nurse swept past them. The wretched woman lifted her key ring to the locked gate, a portal through which Fenella had once been able to race down the stairs cut into the hillside. Fenella silently cursed Argonshire for adding the barrier after her escape last year.

"Your—" Connell's words were cut off by a shout from the lower courtyard, accompanied by the clamor of horses and riders.

Guards manning the upper courtyard portcullis gave answering shouts. Timbers squeaked and chains rattled as the iron-tipped gate slowly rose. Fenella's heart skipped several beats. *Edward!* She wished she could peer into the narrow, roofless, high-walled passage used to access the upper bailey. Intruders would be trapped by the upper gate and met by flaming arrows from the towers. But raising the portcullis indicated friends or clan's men were in the passage.

"That must be Mr. Blackwood." She squeezed Connell's arm. "I wish him to hear your news, as well." *And learn whether he had been aware of your last visit.* "Come, I will introduce you."

As they crossed the upper courtyard together, Fenella resisted the urge to run towards the fortified gate. Uncle Connell led his steed. The beast's ears were pricked to the clatter of horses' hooves in the passage. Fenella kept her gaze trained on the portcullis as dark clouds passed afore the sun, sending shadows that dappled the courtyard.

Mr. Blackwood emerged first, braw and striking in the saddle, his raven hair windblown. His hazel eyes found her immediately, and his lips curved upward in greeting.

Flooded with relief, Fenella returned his smile. Thank Fate her long skirt hid her wobbling legs. She clutched Connell's arm for support.

Edward's gaze shifted to her uncle, and he went still in the saddle. He scanned the courtyard afore urging his horse forward.

"Stop circling us like a wolf sizing up a threat, Mr. Blackwood, so I can introduce you to my uncle, Connell Thorburn."

"I was nae circling, Thaness." Behind him, the last of the men and horses threaded through the gate. Two russet-furred wolfhounds, looking sore footed and thirsty, caught her eye, then melted into the trees.

Connell acknowledged the younger man. "Mr. Blackwood."

"Aye, Connell Thorburn, the family scholar. Fenella speaks highly of you." Blackwood reined in his steed, dismounted, and extended his hand. The two clasped forearms. "What brings you to the Lowlands? Nae trouble, I hope."

"Ten days ago," Connell glanced at Fenella. "I brought word Fenella's mother was gravelly ill."

Fenella gasped. "Does she still live?"

"When I left," Connell said, "but barely."

Fenella's eyes met Blackwood's. "By now," her voice rose to a wail, "Mother may be…" She could nae bring herself to say the word.

Blackwood's eyes darkened. "I vow, I knew naught of this. To whom did you deliver your message?" Blackwood asked Connell.

Her uncle's eyes grew cold. "His lordship. Your father."

"Where is his lordship now?" Blackwood asked Fenella.

"Receiving a messenger from the south in the oriel chamber."

Blackwood faced the nine riders gathered behind him. "Have the grooms see to your horses so you may quickly sup. Do nae stray off! We ride again within the hour."

With nary a grumble, the small contingent hastened to the stable.

To Connell, Blackwood said, "I apologize for my father. Join me, and we shall confront him together."

Connell glanced at Fenella. When she inclined her head, he answered, "Verra well. I thank you."

Both men handed their reins to a waiting stable lad.

"His lordship has ordered me to my chamber," Fenella told Blackwood, heat blooming anew across her cheeks.

"I have nay wish to exclude you." Blackwood raised her hand to his lips and bestowed a kiss upon her scarred palm. "But his lordship may be more forthcoming if you are absent. The choice is yours."

Fenella frowned. "I am nae one to let others fight my battles."

"Sometimes that is how battles are best won," Connell said.

Aye, and now is nae the time to provoke Argonshire's ire. "Verra well. I shall retire to my chamber…for now. But know this…I *will* see my mother."

"We are of like mind," Blackwood assured her. "I will apprise you of the outcome of our meeting as soon as possible. Prepare to journey…in any case." The pledge she saw in his eyes declared his sincerity.

"This way," he told Connell. They scaled the staircase to the hall wherefrom Blackwood led Connell towards the oriel chamber.

Dogged by Calidore, Fenella made her way through the bevy of interconnecting chambers leading to her quarters in the east tower. *What did he mean, "in any case"? Was Edward willing to defy his father should consent nae be granted?* A tinge of hope swelled her heart.

She had to pass through the far corner of Argonshire's spacious quarters, which led to Gavin's generous, unused chamber. From there, the two east tower rooms were accessed. When the errant eldest son returned, Fenella had been warned, she would be moved.

I hope to be living where I belong, Castle Thorburn, long afore then.

Meanwhile, whatever Edward might have in mind should his request be denied, she could nae flee the stronghold without first passing through Argonshire's private quarters. And she could nae escape the grounds without a gate key or means of silently raising the portcullis.

Fenella touched the Thor's hammer she wore on a slender rope of gold around her neck. A thrill shivered her skin as she remembered how Talfryn had magically spirited her from one place to another. Then her thoughts darkened. *He must be dead. Otherwise, he would have appeared to me by now.* 'Twas a fancy to think there had been more than the fellowship of mutual assistance between them. A necklace to remind her of a friendship was all the token symbolized.

Besides, she had vowed to never again betray Edward. She would nae break his heart, nor his trust a second time, for then she would surely lose him. And by his instructions that she ready for travel, she had reason to hope him willing to put *her* interests afore his father's.

Repressing thoughts of Talfryn, Fenella tucked the charm beneath her neckline, then worried her thumb across her scar. Now, more than ever, her fate was in Edward Blackwood's hands.

Chapter Eight
Edward

As Edward opened the door behind the dais to enter the oriel chamber, the enclosed bay serving both as a courtyard viewing platform and a small receiving room, he came face-to-face with the messenger from the south. The stranger's aura of malevolence and secrets shoved against Edward's chest. A chill shivered through him, and the sour taste of evil coated his mouth. He fought the urge to step aside, to escape.

Connell Thorburn eased close, his sword hand ready.

The messenger stared down his nose and coldly assessed them. Thin and balding, the man sniffed the air like a royal catching scent of an unwashed peasant. His haughty air ill matched his clothing, a simple, monk's traveling robe with a rope belt.

Though the man spoke nae a word, the twist of his lips as he strode past sent an unspoken warning: *I will be watching you, Blackwood.*

Edward curled his fists. *What business have you here?* He dreaded the possibilities. The stranger's odor lingered in the doorway, a sour mix of sweat, frankincense, and myrrh. Breathing out the foul stench, Edward shifted his attention to his father. "Your lordship."

Argonshire stood afore the tall bay window. He clutched the shoulder muscles on either side of his neck, his elbows pointed towards the roof. Edward's throat tightened. An image flashed in his mind of the last time he had seen his father thus—when Argonshire had received a letter from Gavin and would nae disclose its contents.

Lord Argonshire lowered his arms but did nae turn to face him. His chin swiveled towards his shoulder, exposing a profile made harsh by time and loathing. "Did you resolve the matter at Innis Chonnel?" he asked without greeting.

"Aye, I did your bidding. But there is an urgent matter afore us. Lord Argonshire, may I present Lady Fenella's uncle, Connell Thorburn."

Connell stepped forward. "Your lordship."

With a disinterested air, Argonshire twisted towards them and arched his brow at Connell. "You have returned," he said flatly.

So, you did know! Edward exchanged a sideways glance with Connell.

"Aye." Connell produced a folded, sealed letter and offered it to his lordship.

Argonshire glared at the missive afore snatching it from Connell's hand. Nae breaking the seal, he asked, "Is she dead yet?"

"My lord!" Edward flinched.

"At the time the letter was written, Máire Rose Thorburn was still alive. Though barely so." Connell's hard tone held little hope Fenella's mother still lived. "I request you allow Lady Fenella to return home with me to say her final farewell."

As if considering the request, Argonshire tilted his head.

"My contingent can be ready within the hour," Edward said.

"Indeed." Argonshire's beady gaze drifted back to Connell. "You brought nay contingent? Are you fearless, or foolish?"

"Would you have opened your gate to a contingent?" Connell challenged.

Edward narrowed his eyes at his father. "How do you know he traveled without—"

Argonshire held up his hand, cutting off Edward's question. "I know everything."

"Then mayhap," Connell said evenly, "you can tell me if Lady Fenella will see her mother alive."

"What assurance do I have the lass will be returned to me afore her wedding on the solstice?" Argonshire demanded, ignoring Connell's impertinence.

"Why nae allow Lady Fenella and your son to marry at Thorburn?" Connell countered.

"They will marry here." Argonshire's tone brooked nay argument.

"We will return afore the solstice," Edward assured his father.

Lord Argonshire tapped the unopened envelope against his palm, flashing the oval abbot's seal. He tossed the missive onto a small chest beside the bay window. "I grant you three weeks."

"Thank you, your lordship," Connell said.

"Three weeks does nae give us much time," Edward said, containing his dismay at his father's guile. "But I thank you for granting our leave."

With a wave, Argonshire dismissed them both. Connell bowed. Once the two men had crossed the threshold to the hall, Argonshire called out, "Edward, a word afore you leave."

Edward cast a fleeting glance at Connell and the guard awaiting them. "Tell Cook to feed our guest and my men," he instructed the guard. "Then send word to Lady Fenella, Derry, and Calidore to pack quickly so we may leave forthwith."

"Just the lasses," Argonshire corrected.

The guard nodded, then led Connell into the hall. Once the door was closed behind them, Edward said, "I could use Calidore. He is a good man."

"Never let too much sympathy build between a guard and the thaness. Take Angus instead. He is good with a sword. And pick five of my men to accompany you."

Perturbed at his father's interference, Edward scowled. "Angus and your men are nae needed." *And they are nae to be trusted.*

Argonshire beckoned Edward close, then leaned in and spoke into his ear. "Take my men and use your time well. Watch for signs your betrothed carries the taint."

Uncertain of his father's true intent, Edward furrowed his brow.

"Any sign she has been touched by the fairies or witches—"

"Lady Fenella does nae associate with fairies nor witches," Edward insisted.

As if he pitied his son's ignorance, Argonshire shook his head. "*Any* sign she has been touched by fairy or witch, marry her. There, where her clan can witness the nuptials."

"Nae here? But only moments ago, you insisted we marry at our stronghold."

"If she is tainted, then best to be done with it afore she turns ten and six, when her powers will become stronger. 'Twill be easier this way."

Dread constricted Edward's chest. "What will be easier?"

"Killing her."

Chapter Nine
A Grave Sorrow

Castle Thorburn, Scottish Highlands

Fenella stood hunched, numb and disbelieving, at the small table beneath the shuttered window in her mother's chamber. She traced her finger down the edge of the open prayer book, barely registering the blue-green decorative border and the red and black text. The entry where she had recorded her mother's death blurred as she blinked back tears.

"I should have been at her side." Fenella swallowed a rough lump of grief, guilt, and rage. The chill in her mother's bedchamber swaddled her, raising gooseflesh. No fire had been lit, and per custom, ashes had not been removed from the hearth. Light sliced in from the door open to the adjoining chamber. A decaying bouquet on the chest containing her mother's garments released a musty odor.

Her cousin Sorley raised his chin and sniffed the air. "We should nae be in here."

"Shall I remove the flowers, my lady?" Derry asked from the doorway.

"Aye. Thank you." Fenella turned and eyed the brown-haired maid. The lass had nae taken well to the long horse ride and had paled further upon arrival at the donjon.

"Of course, my lady." Derry tilted unsteadily with each deliberate step. She retrieved the dead bluebells and hurried out of the chamber. Fenella glimpsed Angus in the connecting chamber afore Derry absentmindedly closed the door, plunging the room into darkness.

"Derry!" Fenella called out.

Angus reopened the door. Their eyes met, his guarded and suspicious. He left the door ajar, then turned his back. Fenella rubbed her arms. Blackwood had given nay explanation as to why Angus had replaced Calidore, nor why some of Argonshire's men had joined the contingent. But his grim expression had told her the changes had nae been his idea.

"Bluidy Blackwoods." Sorley scowled. "If Argonshire had admitted Uncle Connell the first time—"

"I would have arrived while my mother was still alive." *Now I am orphaned.* Pressure welled in her throat. *Father died afore my birth. And now I have been denied a final farewell with my mother.*

She sniffed. "Did they toll the bells upon her death?"

"Aye," Sorley answered, "and at her graveside."

"Good." Fenella's hand trembled as she straightened the leather-bound prayer book. Her mind flashed on Merlin's grimoire. She had glimpsed the small book after the wizard had imprisoned the Lady of the Lake. *Might it contain a way to avenge my mother? If I were schooled in magic...* Her thoughts floated as she wandered to her mother's bed. Salt crunched beneath her boots. *It must have fallen from the night watchers' pockets.* "Who prepared the body?"

Sorley lowered his gaze. "Your gran."

Heat warmed the skin beneath the freckles dotting Fenella's cheeks and nose. "I hope Gran begged Mother's spirit for forgiveness."

"You should ask her."

Fenella snorted. "Gran is dead to me."

"You doona mean that."

Her green eyes met his blue. "'Twas Gran's shove that caused Mother to hit her head."

"Aye, but what if the Lady of the Lake's spell had guided the old woman's hand?"

Aye. What if? Fenella released a long breath. Her gaze drifted to her mother's small mirror atop the garment chest. Servants had shrouded it, along with all the mirrors in the castle. None of her mother's essence remained. The ache in her heart expanded.

"She fell over a year ago," Sorley reminded Fenella. "We canna be sure Gran's shove caused her death."

"In her letters—" Swallowing back tears, she realized there would be nay more letters. "Mother described…" her voice cracked, "…terrible headaches. Profound pain kept her abed." She paused to regain her calm. "So, if naught else, I blame Gran for Mother's suffering."

"Da agrees with you."

"As he should." Fenella imagined Mother's other brother Dunstan agreed, as well. She glanced at her mother's empty bed. A fresh wave of loneliness struck her, and she shut her eyes against the sudden crush of memories. The scar on her palm throbbed like a wounded heart, and she scrubbed the key-shaped brand with her thumb as if to wipe it from her flesh, her thoughts… her very life.

Sorley stayed her hand. "You need the scar. Proof of who you are, remember?"

"How could I forget?" Bitterness edged her words. The searing pain from Gran's fiery brand flared anew. She sucked air between her teeth, reliving the scorching of her flesh. Then, with a waving motion, she stirred the chilly air, pushing it against her palm.

Sorley clasped her wrist, returning her to the present.

Color bled back into a world gone gray.

Fenella gathered her wits. "I will marry Edward Blackwood on Saint John's Day," she continued in a whisper, "the day of my birth." *The solstice.*

Sorley swept his gaze to the chamber door. "Here?"

Fenella shook her head. "His lordship requires we hold the ceremony at the Blackwood donjon."

"Why dinna he force the marriage after his thieves stole you?"

Her chest constricted; she glanced at the door. She sensed Angus listening out of sight, just beyond the open crack. "Mr. Blackwood intervened," she whispered in Sorley's ear. "He asked that I have time to adjust. We grew close, 'til…"

"'Til you followed me into Fairy." Sorley whispered.

Fenella sighed. *Edward appears to love me again. But trust me?* The ache upon her heart worsened. She could nae blame Edward for being wary. Nae after she had betrayed his trust to rescue Sorley.

"I am surprised Argonshire agreed to a delay."

"He had his reasons." His agreement to such a long postponement worried her. Lord Argonshire's stronghold was riddled with spies watching for signs she carried "the taint." A lingering fear prickled her scalp. The longer her marriage was delayed, the higher the chance her magic would be discovered.

Fenella eyed Sorley's wild mane, identical to hers with its bright mix of red, copper, and ginger strands. She hooked her arm through his and led him away from the door and Angus's attentive ears. "You know what they say aboot us."

Sorley scowled. “Aye. Fairies. Witches. The fools. ’Tis in our blood.”

“Mayhap. Lord Argonshire fears I will taint his precious bloodline. But he covets our donjon because it gives him a northern stronghold.”

“So, he awaits your sixteenth birthday.” Sorley lowered his voice further. “My abilities showed when I was a wee lad. If you were able to shift shape, ’twould have happened by now.”

“But what if the taint reveals itself in other ways?” she whispered.

Sorley arched his brows. He leaned closer. “Has it?”

“While I was in Fairy…” She hesitated. ’Twas blasphemous to think, much less say aloud.

“Go on.”

Her gaze darted to the door. “Talfryn believes I performed magic.”

“Truly?”

“Shh!” She clasped her hand over his mouth. Sorley narrowed his eyes, and she lowered her hand. “Fairy altered me.”

“Then you must take care.”

Fenella nodded. “Argonshire suspects, but he bides his time. My wardship will end when I marry. ’Til then, the earl controls my fortune and our clan’s future.”

“What happens after you marry?”

“If the Blackwoods doona kill me?”

Sorley drew back. “Cousin!”

“Argonshire is ruthless. He caused his own son to go lame! My safety is nae assured…which is why Argonshire must never discover my magic, and my marriage to Blackwood must be a love match.” *If our bond is stronger than Edward’s allegiance to his father…*

"A love match?" Sorley scoffed.

"Aye." Fenella gripped his arm. "I beseech you. Befriend Edward Blackwood. You further neither your cause nor mine by demonizing him. 'Twas Argonshire who stole me, nae his son."

"All Blackwoods are alike." Sorley curled his upper lip into a snarl.

"Nay! After I was stolen, Edward Blackwood nursed me back from death's door. And I am here now because of him."

Sorley expelled a quick breath through his nose. "What aboot Talfryn?"

"Nae a word."

"You doona think—"

"Talfryn was deathly ill when we last saw him." She fingered the small Norse charm she wore around her neck. "I fear he did not recover. Otherwise, he would have shown himself by now."

"You wear his charm." Sorley jutted his chin towards her Thor's hammer.

Muffled male voices sounded in the adjoining room. Sorley held up a warning finger, then angled his head towards the noise. After a moment he said, "You still hold out hope."

"I *hope* my marriage to Blackwood will safeguard our clans and free us from his father."

"And Talfryn?"

"I pray he survived, but I have mourned his loss." *I cannot think of him.* Nae now. "He risked his life to help us." *Another debt owed.* She wondered if her alliance with the Blessed Court had ended when Talfryn…passed. She tucked the slender chain beneath the neckline of her léine. The charm slid over her heart.

Sorley eyed her. "Why did you summon the daft monk, Eilif, to Argonshire?"

"For spiritual advice…and he is nae daft. Besides, he returned to Thorburn months ago, as you well know."

"Did he also teach you the Norse language?"

She raised her chin. "I now have limited command of the Norse tongue."

"And what need have you of this language?" He paused for affect. "Ah, Talfryn is half Norse."

"He is also half fairy, and I have never heard him speak a word of Norse."

The voices outside the door faded.

"We shall never see Talfryn again," she whispered. Exhaustion and grief made heavy her bones, and she felt older than her nearly sixteen years.

Sorley fixed his keen gaze upon her. "Do you *want* to marry Blackwood?"

A small yearning stirred her core. Edward's many kindnesses, and the endearing way he tucked his errant hair behind his ears, spooled through her mind. "I believe so." Then she remembered Argonshire, and her mood darkened. "Edward suffered unforgiveable consequences because I ran off. 'Tis a wonder he wishes to go forward with the marriage."

"Blackwood has as much to gain from the match as you do," Sorley reminded her.

Her heart twinged. *Does Edward seek love, or does he wish for power and freedom from the earl? Mayhap our fates and desires are entwined.*

Sorley's gaze dropped to the floor. "I am sorry. You jeopardized your future when you followed me into Fairy."

"You couldna have foreseen what was to come." With a pang, she remembered how each obstacle she and Talfryn had encountered had increased their bond during their search for Sorley. A year ago, she had returned to captivity more out of a sense of duty to protect her clan than… Guilt and shame heated her cheeks. "We reveal who we are by the promises we keep…or break. I willna break my promise, nor Blackwood's heart, again."

"Understood." Sorley's eyes met hers.

"There must be whispers," Fenella fretted. "Fourteen months, and I am neither free nor married."

"Announce your wedding plans."

Fenella shook her head. "The time is nae right." She retrieved the prayer book. Her body craved sleep; her mind begged a reprieve. But Argonshire was forcing Blackwood to set a punishing pace so she would have neither.

"You look close to collapse," Sorley said.

"A few gulps of fresh Highland air will revive me."

"Then we shall descend to the bailey and find you some."

"Aye." She followed Sorley to the chamber door. Pivoting at the threshold, she whispered, "However long it takes, I promise Lord Argonshire will rue the day he crossed Clans Thorburn and Rose."

Chapter Ten
No Way In

The Woods Outside Castle Thorburn

Talfryn materialized in a forest, breathing in the loamy scent of damp earth. Overhead, a mossy oak canopy barred most of the sunlight, shrouding him in shade. A cool ground mist swirled against his legs; his toes curled within his boots.

Puffball tugged on his ear. "Did you see them?"

"Who?" A branch above him shook as a red squirrel scurried to a higher limb.

"Fenella and Sorley!"

"Where?" Talfryn's stomach jumped.

Puffball stuck out her thin apple-green arm and pointed to the brooding stone tower. "Atop those stairs."

"Let me see." Vendel skirted a juniper that blocked his view, then eyed the tall rectangular donjon. "They're gone." He faced Talfryn. "Too much Cold Iron… The place will kill ye."

Nettle, perched on Vendel's head, clutched two clumps of the dwarf's sandy hair. The pixie's long pointed ears twitched. "Now what?"

"Any of you know how human time compares to Fairy time?" Talfryn asked.

The pixies shook their heads.

Vendel burrowed his hand beneath his padded tunic and withdrew a palm-sized pewter pickaxe from what Talfryn surmised was a secret pocket. The fragile-looking charm gleamed.

Dwarf forged. As Talfryn drew closer, he noticed an apparent seam along the length of the handle. Vendel twisted the little axe blade. The handle seam opened with a faint whir, and a crystal disc slid out.

"Ooh! What is it?" Puffball asked.

"A Luminositor," Vendel replied. "One of my wife's ancestors invented it. The device reveals the phases of the moon."

"I didn't know you were married," Talfryn said.

Nettle leaned over Vendel's forehead. "I didn't know there were female dwarfs."

Vendel brushed away the pixie. "Mab's Court has nay knowledge of her. Vow you will safeguard our secret."

Puzzled, Talfryn vowed. The pixies bobbed their heads.

"Talfryn's winged horse is named Luminous," Puffball said.

"She glows with starlight," Talfryn explained.

"The Luminositor measures the brightness of the moon to determine the moon's phase." Vendel studied the dark crescent to the left of a gleaming second quarter moon. "The moon in Fairy has not changed since last night. Guess time travel didn't affect the Luminositor."

"Let's examine the device tonight when the moon rises in the human world." Feeling slightly less anxious, Talfryn clasped the dwarf's shoulder. "You and Nettle stay here. Warble like dippers if Fenella or Sorley reappear. Puffball and I will search for woodland fay. They might be able to help."

With a counter twist, Vendel retracted the enchanted disc, then hid the device.

"What's that noise?" Puffball angled her head.

"Dog!" Nettle shrieked, leaping into the nearest tree.

"Quick! Into the oak!" Talfryn hoisted Vendel onto a low, thick branch.

Puffball dove for Talfryn's pocket. "It is behind us!"

As Talfryn glanced over his shoulder, a ratter, teeth bared, launched its small body through the air.

"Do something!" Puffball squealed.

In a burst of white light, Talfryn vanished. Seconds later, he reappeared in a pine-studded part of the woods. The ratter's barks sounded in the distance.

"Do you think Vendel and Nettle will be safe?" Talfryn's breath formed small clouds in the cool Highland air.

Puffball popped out of his pocket. "Aye, unless someone from the castle investigates the barking."

Talfryn scanned the forest for a green glade, a riverbank, or a bend in the path, any spot where the fay might gather. He sniffed the air. "Elves are about."

"Which kind of elves?" Puffball whispered. "Woodland, warrior, house elves, or the beautiful kind?"

Something hard and pointed hit Talfryn's head and bounced off. Puffball scurried after the object and plucked it from the thick layer of pine needles on the ground. "Just an acorn." She tossed it on the ground.

A second acorn struck Talfryn's shoulder. He squatted and retrieved it, rolling the smooth nut between his fingertips. "Do acorns fall from pine trees?"

"Ah, nay." Puffball pressed her cream skullcap to her head and tilted her chin up. "I see a blue-gray cuckoo," she whispered through clenched teeth, her thin lips barely moving. "Possibly a small woodland elf."

Talfryn's eyes narrowed. Faster than a blink, he disappeared, then reappeared on the high limb beside the cuckoo. He lunged afore the bird could escape.

"Release me!" the cuckoo screeched.

"Nay, elf." Talfryn held fast to the bird's short yellow legs. "Not 'til you assume your true form."

The yellow-eyed cuckoo fluffed out its drab blue-gray feathers, then pumped its body up and down.

"Oh, for Mab's sake. Are you trying to scare me?" Talfryn asked. "You look ridiculous."

With a hiccup, the cuckoo transformed into a sprite half the size of a tree spirit. Leaves and moss had been woven together with spider's web to create her clothes. She batted her long, silver-green lichen lashes. "I'm not an elf. I'm a moss maiden," she said in a high-pitched voice. "Unhand me."

When Talfryn noticed the gleam in the female's dark green eyes, he tightened his grip. *I knew woodland elves were about!* "Moss maidens are found along walls or at the base of trees, not high in the branches. Reveal yourself!"

"I told ye. I'm a moss maiden."

"Then explain your pointed ears," Talfryn swept the creature upside down. "And how you obtained a stash of acorns."

Nuts tumbled from her hand. "Let me down!"

"We have no time for this!" Puffball called from below. "Drop the cudweed on her head. That will make her talk!"

Pointedly, Talfryn peered down at the ground. "It is a long way."

"Just put me back on the branch!"

Gladly, Talfryn obliged.

The being glared at Puffball, then shifted into a frumpy elf dressed in a smudged tunic and wrinkled leggings. "These aren't yer woods, Trooping Fairy. They belong to the Lady of the Forest."

"I have no designs on your mistress's woods," Talfryn said. "I seek the Lady of the Castle, the Thaness of Thorburn."

"*Ye* seek the thaness?" The elf cackled and fanned herself with her hand. "Now, that's passing strange. Did ye *see* the stronghold? Did ye *smell* it? It reeks of Cold Iron. Ye would never advance past the iron gate."

Talfryn folded his arms across his chest. "Then I'll wait for her to emerge."

The elf rubbed her nose. "If ye wait here, ye will miss her. The drawbridge is on the other side of the castle. A manor elf told me the thaness leaves today for Argonshire."

"Mab's tears!" Talfryn scrambled to the ground. "Puffball!"

The pixie leapt into his hand. "But the dog—"

"We can't miss Fenella!" In a heartbeat they were back with Vendel and Nettle high in the oak.

"Away with ye!" Vendel told the ratter.

Rather than obey the command, the tenacious beast launched itself against the tree trunk, barking and twisting as it jumped through the air.

"Guards are looking this way," Nettle warned.

Talfryn glanced at the stronghold. Six stern-faced guards stared over the parapet wall and pointed in their direction.

"Oh, for Mab's sake!" Puffball hurled a pinch of pixie dust. The ratter fell, mid bark, onto the loamy ground and quietly snored.

Vendel glared at Nettle. "Why didn't *ye* think of that?"

Nettle stuck out his tongue.

"Shh!" Talfryn glared at them. "Someone is coming."

He pushed aside a leafy branch and peered down through the tree limbs. Two human males clomped towards them, crushing twigs, lichen, and honeysuckle beneath their laced deerskin boots. As they drew uncomfortably close, Talfryn noticed the strapping younger Scot walked with one hand pressed against his thigh, as if his leg pained him. The older man, his face wily and his eyes filled with secrets and schemes, carefully kept pace. Both wore pleated yellow shirts that hung to their knees and rectangular wool mantles to chase away the cold. The coverings, black as a night forest during the dark moon, were edged in blood red, which reminded him of the murderous Redcaps he had seen in the Borders. Remembering the malevolent goblins that soaked their caps in the blood of those they had killed, he shuddered.

Talfryn guessed the younger Scot was close to his own age. As the two men neared, Talfryn inhaled Cold Iron. His throat tightened. He drew back as much as he could without rustling the branches and fingered his father's thunderstone charm.

"I see neither the dog nor the cause of the barking," the older man said.

"The wretched beast must have chased a deer." The younger Scot hitched the bow slung over his shoulder to a higher position and said in a low voice, "No matter. The dog was a ruse. I wished a word with you alone aboot the journey ahead."

"Do you fear an attack?"

The younger man scanned the trees. Talfryn held his breath.

"Nothing so obvious."

Talfryn noted the thinly veiled appraisal in the older man's eyes.

"The garrison will be on alert," the older man said.

"I will nae tolerate mistakes, Angus. Do nae be in your cups nor too quick with your sword."

"There will be nay trouble, Mr. Blackwood."

Mr. Blackwood! A flash of realization jolted Talfryn. *Edward Blackwood, the earl's son? Fenella's betrothed?*

Angus adjusted his mantle. "We will make sure your betrothed is delivered to Argonshire. She will nae be able to delay your marriage."

Blackwood's voice turned to Cold Iron. "Speak of the thaness with respect."

"I meant nay disregard, Mr. Blackwood."

Edward glared.

"Will your brothers return for your wedding?" Angus asked in an obvious attempt to change the conversation.

"There has been nay word."

"Shame. You were so close to Hamish afore he was sent away."

"Hamish was nae *sent* away. He joined the monks when he was Lady Fenella's age. And Gavin has long been an adventurer. I barely remember him."

An uneasy silence fell between them.

Blackwood scrubbed a lock of his long dark hair behind his ear. "I sensed an unease among his lordship's men on our journey here. What whispers concerning my upcoming marriage swill among them?"

Angus glanced away. "Lord Argonshire fears a fairy or sorceress touched Lady Fenella at birth." Angus cleared his throat. "She *was* born on the summer solstice. Och, her gran probably started the tales to chase away suitors." Angus shook his head. "Anyone who would brand her own granddaughter…"

"Would be banished if I had any say." Blackwood glanced towards the stronghold. "And the whispers are nae true."

Angus dropped his gaze, then proffered, "Some believe Lady Fenella and her cousin fell down a fairy mound after she was brought to Argonshire for safekeeping."

Blackwood rubbed his leg. "She dinna."

"I have heard sorcery or fairy taint reveals itself by the sixteenth year. If she shows nay sign of it by her birthday, then she should be safe. If she does..." Angus spit on the ground.

Eyes narrowed, Blackwood slid his hand to his sword hilt. "Taint or nay taint, I will nae abide any harm coming to her. You will do well to remember that."

"Of course." Clearly rattled, Angus dropped his gaze. "As you wish." Then he glanced at the bracken. "Is that Lady Fenella's dog? What ails the beast?"

"How odd." Blackwood slid the bow off his shoulder and pulled an arrow from the quiver on his back. "What ails it, indeed?"

As Blackwood nocked the arrow into the bowstring and pointed it at the spell struck dog, Talfryn tensed. Blackwood hesitated; his eyes trained down the shaft. Abruptly, he swiveled, arcing the bow and arrow towards Angus.

Talfryn's heart shuddered to a halt. He clasped his hand over Vendel's mouth to silence the dwarf's gasp.

"Lady Fenella will soon be my wife, Angus. Should *any* mischief befall her, the bonds of kinship will nae spare those involved. Do I make myself clear?"

The lump in Angus's throat bobbed up and down. "Aye."

"Good." Blackwood returned the arrow to his quiver. "How odd that the beast doona wake."

"Verra odd."

Blackwood hesitated, and Talfryn thought the Scot might gather up the sleeping dog. Instead, Edward rubbed his lame leg, then glanced at the stronghold. "Come. We have a contingent to feed afore we move out."

As they tromped away, Talfryn and Vendel released long breaths.

Puffball fanned her face with her hand. "How dashing, for a human!"

"Fenella plans to marry him?" Nettle asked.

"Not by her choosing." Talfryn's skin buzzed with restless magic.

"He defended Fenella and warned he would protect her," Puffball argued.

"He's the son of her kidnapper. Why should we trust him?"

"He convinced me." Puffball crossed her spindly arms over her chest.

"I hope your faith is well placed." Talfryn held out his hand for the others to clasp. "To the drawbridge."

They reappeared on the eastern side of the castle, landing in a grove close to the moat that ran beneath the drawbridge. Vendel ducked behind a large juniper while Puffball settled noisily into its fragrant branches. Nettle raced up a nearby wych elm.

Talfryn paced, his gaze fixed on the keep. He withdrew the thunderstone from beneath his neckline and traced his thumb over the smooth crystal, avoiding the thorny hook. Mab's warning repeated inside his head. *"You better hope Lady Fenella's heart belongs to you."*

What if Fenella's heart belongs to Edward Blackwood? Sweat dampened the front of his ridiculous saffron shirt.

"Regarding Edward Blackwood—"

"I do not wish to discuss him." Talfryn glowered at Puffball.

Nettle called down in a loud whisper from the wych elm. "I thought Talfryn and Fenella were just close friends."

Puffball shrugged her tiny shoulders.

"The lad is still sorting it out," Vendel explained as he emerged from behind the juniper.

"What about Lily?" Nettle persisted.

"Enough!" Talfryn fairly shouted.

As though she could not stop herself, Puffball said, "Once that human marries Fenella, he becomes part of her clan, to which you've sworn allegiance."

"All the more reason to prevent their marriage 'til we're certain Blackwood is unlike his father."

"One step at a time, lad," Vendel told Talfryn. "First, we retrieve yer thaness."

Chapter Eleven
A Second Chance

Fenella's Day of Departure
Castle Thorburn

Fenella insisted an empty space be left on the bench at the high table to symbolize her mother, Máire Rose Thorburn. She claimed the area beside it and laced her hands together to stem their trembling. Seated on her left, she saw that Edward kept his sword close and his dirk within reach. *Does he expect my uncles will burst forth from a secret passage and slaughter him for the sins of his father?*

As her spirit hovered overhead connected by a silver thread, Fenella blinked back tears. All color faded from her sight, plunging the hall into shades of ash 'til a sole spot of color drew her gaze—the blood-red diagonal slash on the Blackwood banner.

Her spirit dove back into her. 'Twas as though the standard's symbol sneered down at her, reminding her of how much Argonshire had stolen. Her stronghold. Her freedom. Her life!

How dare that cloth hang possessively in my hall!

A tempest erupted inside Fenella's goblet, rattling the vessel and sloshing wine over the rim. Stunned, she grabbed the stem and the uproar stopped. Color bled back into the hall as she glanced about. No one seemed to have noticed the odd occurrence.

Fenella took a calming sip of the creamed turnip and parsnip soup afore her. *I must eat. Stay strong. Control my fury.* She managed two spoonfuls.

An undercurrent of tension accompanied the subsequent roast beef and leeks with almonds. Thorburn guards ate in silence, casting curious, sometimes pitying stares at her and murderous glances at the interlopers who had stolen their thaness and held her against her will.

"Do you think 'twill ever be thus?" Blackwood's musing interrupted her construing. He smoothed his hand across the tablecloth, resting near hers. "Hostile, joyless?"

"'Tis hard to imagine otherwise." Fenella met his worried gaze. *Remember your goal.* She exhaled her fury; refocused.

"I've heard a happy marriage makes for a happy hall." The gold flecks in his hazel eyes caught the candlelight. "But under the present circumstances…"

She lowered her gaze and softened her tone. "With Mother newly buried…"

Silence stretched between them. Fenella fingered her knife.

Blackwood studied the assembled crowd. "Your gran failed to join us. Has she taken ill?"

"I dinna invite her."

Blackwood glanced at Fenella's scarred hand. "The woman has much to answer for. Still, if we are to marry—"

Fenella's heart faltered. "If?"

Their eyes met. Her stomach flipped. Blackwood slid his hand closer and hooked his finger over hers. Fenella's breath caught.

"When," he corrected.

The tightness in her shoulders eased a fraction. "So, you doona want to pass me off to one of your brothers?"

Blackwood's fingers tensed, and his forehead wrinkled with his apparent distress. "Is that your wish?"

She touched his knee, like a song thrush briefly alighting on an apple branch afore being frightened away. "Of course, nae."

His handsome features relaxed. The warm flecks in his eyes brightened.

Fenella leaned forward and gazed at the men and women quietly eating. Fourteen months ago, she had known the stories of every man, woman, lass, and lad seated afore her. But now…

Argonshire's men lorded over the closest table, where once her guards and the king's men had sat. Edward's contingent commanded the long trestle table across from them. Her kinsmen claimed the lesser tables, those farther from the hearth and high table. She spotted Balfour, her most trusted Thorburn guard, and a few of the other men who had been with her on the ramparts the night the Blackwoods had stolen her. To a man, they dropped their gazes.

'Twas nae your fault, Fenella wanted to tell them. *'Twas the Lady of the Lake's spell cast forward in time that prevented you from protecting me.*

Balfour glanced up, an apologetic expression on his weathered face. Fenella inclined her head.

"A kinsman?" Blackwood asked.

"Aye." It seemed imprudent to single out her guards. "Since I was nae at the gravesite…"

Blackwood covered her hand with his. "I am sorry you could nae mourn with your clan."

"I thank you." His warmth spread up her arm, dispelling some of the emptiness within her. The room stilled, and she sensed all eyes upon them. A blush warmed her cheeks.

Lord Argonshire's men crossed their arms and jutted their chins. Some smirked as though their future tenure at the stronghold had been confirmed by Blackwood's gesture.

Do nae become too comfortable in my Great Hall. You will be the first to leave, Fenella swore her resolve in silence. She glanced sideways at Sorley seated to her far right. Eyeing Argonshire's men, his upper lip curled into a snarl.

Many of Edward's men raised their brows and paused over their trenchers. *They must wonder if the castle will soon be their new home.* Fenella glanced at Blackwood. He rubbed his thigh as if the year-old dog bite on his calf pained him still.

A fresh wave of guilt swamped her. She rotated her hand. Skin-to-skin their palms met, the thrill, a lightning strike.

Blackwood's lips parted. His hand stilled on his thigh, and he shifted closer. "You have nae eaten," he observed.

She shook her head. "Grief and nerves have robbed me of hunger." Thoughts of her mother welled within her. *I will nae shed tears, nae in front of Argonshire's men. I must nae appear weak.* She lowered her chin and mumbled, "What must my people think of me?"

"They think you brave and strong."

"Mayhap… But, chaste?"

"Ah." Blackwood squeezed her hand, then cleared his throat and rose.

Fenella's heart stammered. *Will he abandon me to my shame and grief?*

Instead of departing, Blackwood shifted his weight to his left leg, then addressed the hall. "For those who have nae met me, I am Edward Blackwood."

Faint grumblings sounded from the Roses and Thorburns seated in the back.

Angus glared and rose to his feet.

Blackwood gestured for him to sit.

"Most of you know my betrothed, Lady Fenella, the virtuous Thaness of Thorburn." When Sorley led the applause, Edward bowed and raised Fenella's hand to his lips.

He released her hand and pressed his palm to his chest. "Our hearts are heavy over the passing of Máire Rose Thorburn. Fate prevented me from meeting her, just as Fate seized my mother as she gave birth to me." He glanced down at Fenella. "Your loss is my loss," he whispered.

Words escaped her.

Edward again addressed the hall. "'Twould relieve Lady Fenella to know her mother had a proper wake. Did the night watchers entertain you with a good duel?"

"Aye!" roared the men in the back.

"Did the ale flow?"

The hall shook with table pounding.

"Did the night watchers light one candle from another?" Blackwood called out.

Fenella crinkled her brow.

One of her former guards broke the sudden silence. "Nay!"

"'Tis forbidden!" the men spoke as one.

"Good." Blackwood clasped Fenella's shoulder. Warmth flooded her. "'Tis the same in the Highlands as 'tis in the Lowlands."

Argonshire's men and Blackwood's men joined in the shouting and table pounding. Fenella gaped.

Blackwood released her and leaned towards the crowd. "Did the castle provide pancakes the morning after?"

"Aye!" came the shouts.

"Another tradition we have in common!" Blackwood smiled down at Fenella.

She scanned the crowd. *He is uniting them.* Her heart expanded.

Blackwood gestured to Sorley, then shook his head. "A brave lad, your Sorley Rose. Tell me—nay—warn me. Are all Roses thus?"

Ayes and laughter wafted from the clan Rose table. "You have nay idea!" one shouted, followed by sideways glances, smirks, and more laughter.

Sorley surged to his feet. "Da said Fenella's mother had a brave heart. She loved to roam as a wee lass." Sorley lowered his voice and directed his next words to her. "Da will be sorry to learn he missed you. If only you could stay a day longer…"

"I wish 'twere so, but I must depart forthwith."

Balfour caught Fenella's eye and called out, "Máire Rose Thorburn bravely sacrificed for the good of the clan."

Fenella was bolstered by and grateful for the meaning behind his words. She remembered her parents' forced marriage, decreed by the king to broker peace between Clan Rose and Clan Thorburn. Her clan remembered too.

She stood. "Your kind words warm my heart. And whilst I was nae in attendance, it comforts me to know my clan's men and women took the proper actions in my stead. I am certain your prayers eased my mother's heavenly journey."

Then Blackwood raised his chalice. "May the Light Perpetual shine upon Máire Rose Thorburn."

All raised their glasses and echoed Blackwood's words. As the hall quieted, Blackwood and Fenella regained their seats. His lips separated into a bemused smile. "You are staring. Have I splattered soup on my face?"

She shook her head. "Nay. Your countenance is perfect."

****

Talfryn paced within an elm copse across the river from the stronghold. Imagining Edward Blackwood and Fenella together inside the Cold Iron-infested castle, he flexed his fingers to dispel his frustration. *What is taking them so long?* Is Fenella safe? "Where is she?" he muttered. Picturing his father tethered to Cold Iron whirled his stomach.

"Fenella would hasten if she knew Beacon was in danger," Puffball said.

"Shame we can't convey word to her." Vendel studied the castle.

"The elf probably lied to you." Puffball frowned. "It is already past midday."

Vendel flinched. "The ground is trembling. Horses."

Talfryn and Vendel concealed themselves behind an elm while Puffball raced up the trunk and huddled within its leaves next to Nettle.

"Keep quiet," Talfryn ordered.

A dozen riders burst from the stronghold and onto the drawbridge. A dreadful jangle of Cold Iron bridle rings rang above the clatter of Cold Iron-shod hooves on the wooden bridge.

In the midst of the fierce-faced, broad shouldered men, Talfryn spied a slender figure. His breath caught. He couldn't see her face clearly, but the long coil of flame-red hair assured him. *Fenella.*

Chapter Twelve
The Encounter

Fenella tightened her grip on the reins. The leather cutting into her palms helped root her in the present. But the crush of Blackwood guards, and the smell of sweat, leather, men, and horses stirred unwelcome memories of her capture, memories she thought she had buried so deeply they could nae resurface. Derry's death grip around her waist was nay help. She forced herself to release a long breath. *I am nae being stolen.*

A rider in front of her leaned to check his scabbard, offering her a glimpse of her betrothed. Blackwood cut a braw figure in the saddle with his warm mantle wrapped around his straight back and his muscular thighs bare despite the cold. Fenella took heart in the way Blackwood's confidence rose away from his father. Even the most seasoned warriors in the contingent obeyed his orders.

Her mind calmed.

Ahead, a horse whinnied. Blackwood raised his hand, signaling the guards to halt. Fenella stood in her stirrups to peer through the shifting mass of men and horses. But she could nae see the reason they had stopped.

"Madame," Blackwood said. "Remove yourself and your horse to safety. You could have been seriously harmed had I nae spotted you in time."

"But you did, and I am unharmed," said a quivering female voice. "I have brought the thaness a gift in memory of her departed mother and in celebration of her upcoming wedding."

A gift? Not recognizing the voice, Fenella urged her palfrey forward. Angus blocked her way. With his sword raised, he reined his horse in a circle, his shrewd gaze searching the trees for trouble.

"Surely you do nae intend to give away your beautiful mare?" Blackwood asked, suspicion creeping into his voice.

"Aye, your lordship. A lesser gift would insult the memory of Máire Rose Thorburn and would not be proper tribute to your nuptials."

She sounds like a foreigner, yet she claims to know my mother and me.

Derry slid off Fenella's horse, and in her curious listing gait, threaded through the riders, flailing her hands as if to ward off any steed that shifted too close.

"Derry, wait!" Fenella called out, again urging her horse forward.

"It could be a trap," Angus warned in a low voice.

Fenella's senses heightened.

"A pure white horse! My lady, come see!" Derry called over her shoulder.

A white horse? Nay. Oh, nay. Dread rooted her to the saddle. Without urging, Fenella's horse broke past Angus then wove through the contingent, carrying her within sight of Derry and the white palfrey. Instead of leather and Cold Iron, the horse wore Beltane ribbons for reins and a fine woven tapestry as a saddle.

"So pretty!" Derry clapped her hands together.

The snowy horse turned its head and met Fenella's gaze. An otherworldly gleam shined brightly in the palfrey's eye.

Fenella's breath caught.

Its head slightly angled, the fairy horse nodded once.

The mare realizes I know what it is. She glanced about, half expecting to see the Lady of the Lake, or Glynis, or...*Talfryn?* Returning her focus to Derry, her stomach tumbled when the daft lass nuzzled the horse as if spellbound. "Derry—"

"Mayhap..." The old woman's pale green eyes met Fenella's.

Familiarity zinged through Fenella. *I've seen those eyes afore, but where? When?* She flashed on her meeting with Viviane, the Lady of the Lake, but could think of nay elder at the Beltane feast. Something about the woman's appearance did nae match with the memory hiding just beyond Fenella's grasp.

"Mayhap," the stranger repeated, her glance shifting to Blackwood, "you would prefer the thaness's maid ride the mare first. Then you would know the horse is safe, a mere gift, nae a trap."

Derry, her hands clasped below her chin, threw Blackwood a pleading look.

"Your offer is most kind," Fenella said from atop her horse, "but..." Derry's expression gave her pause. Blackwood's men had formed a circle around them, half facing the crone, the others facing outward, their hands on their sword hilts. Derry whispered in the fairy horse's ear. The mare's ear twitched, then it shook its head.

"Lady Fenella?" Blackwood said, indicating the choice was hers.

This can nae end well. She met the old woman's pale green eyes again and saw nay malice, only hope. "How will you return home, if I accept your gift?"

The carlin's eyes brightened. "'Tis a short walk to the village. I have nay need of a mount."

Fenella nudged her horse closer. "Have we met afore?"

"Nay, my lady. Though I saw you and your cousin once at a

festival." She handed Derry the palfrey's ribbon reins.

A festival? Once more the memory of Avalon rushed afore her. "Verra well. I thank you for your generous gift. Derry, can you manage?"

"Aye, my lady!" Derry squealed in delight.

As Derry mounted, Blackwood gave a command, and the guards reined their horses back into riding formation, pushing Fenella into the center. Once astride the mare, Derry was absorbed into the throng. Fenella noticed with some unease the white horse stood out among the Blackwoods' dark steeds, like a fairy lantern on a moonless night.

Angus positioned his horse next to Fenella's. Then as one, the group surged forward.

Twisting in the saddle, Fenella called out, "Does the horse have a name?"

"Starlight!" The old woman's appearance wavered. Her lined face smoothed into the countenance of a younger woman. And for a fleeting moment, her gray hair brightened to the colors of autumn leaves.

Fenella blinked, and the mysterious woman vanished, leaving behind nay trace save three raven feathers scattered on the cold summer ground.

Chapter Thirteen
Another on the Road

The woman had thrown a glamour! The glimpse had been too quick, but long enough to be certain she was nae the Lady of the Lake. *Was she a fairy? A sorceress?*

"Morgan le Fay!" Derry mumbled. Wonder tinged her voice, and for a moment, her eyes appeared to tilt up.

"Derry?" Fenella angled her head.

As though suddenly aware of her surroundings, Derry clamped her hand over her mouth, and her eyes appeared normal once more.

Had the shape-shifter bewitched Derry? And why gift me the fairy horse? What might it portend?

Fenella tore her gaze from her maid and focused on the back of Blackwood's head and the solid breadth of his shoulders. She sat straighter in the saddle. Her future rode afore her. Freedom. Family. And if her plan to marry Edward for love succeeded, she could seize control of her thanedom. *I will nae jeopardize my plan. I will nae be lured back into Fairy.*

After about a league, Fenella relaxed her vigilance. *Mayhap the snowy mare had merely been a wondrous gift, nae an ill omen. But why come to me now? Why would she…* A commotion ahead interrupted her thoughts.

From the front of the contingent, Blackwood raised his hand, signaling the men to halt. "What have we here?" Bemusement and wariness edged his tone, rattling Fenella's brittle nerves.

Hoarse as though little used, a gravelly voice said, "I am Vendel Fairheart, also known as Vendel the Dwarf. I have a token for the thaness."

A dwarf? Fenella urged her mare forward.

"Do you seek the lady's favor?" Blackwood inched his horse closer to the dwarf. "Surely you have heard she is betrothed to Edward Blackwood, son of the Earl of Argonshire."

"Aye, I did hear his lordship's father had stolen Lady Fenella against her will."

The ring of metal pierced the air as Blackwood drew his sword. "Fool! The thaness was removed for safekeeping."

"I can speak for myself." Fenella dismounted and shouldered her way through the guards. There, standing on the side of the road beneath a full-grown hawthorn tree that had nae existed afore, stood a determined-looking dwarf. Her gaze traveled down his flat ruddy face to the yellow flowers clutched in his stubby hands.

Her stomach whooshed as she recalled the dried tansies hanging from the ceiling in Talfryn's Fairy home, and the pressed flower he had given her. "Dear Vendel," she spoke as though the stranger were an old friend. "You brought me tansies!"

Vendel flashed her a keen glance. "Nettles would sting. Puffballs might make ye sneeze. But tansies seemed just right for the Thaness of Thorburn." His gaze swept her yellow léine and the scarlet plaid she wore as a mantle. "They match yer, uh, tunic."

"So true." Bending to take the flowers, Fenella asked under her breath, "Is Talfryn alive?"

"Aye, milady."

Her grasp tightened, breaking one of the stalks.

"Very thoughtful of you, Dwarf," Blackwood said from atop his horse. "Thaness…"

"We must take our leave," Fenella apologized. "I thank you for brightening my day."

"Ye are most welcome, my lady." Vendel's intense stare belied his casual tone. "Mayhap ye will encounter more tansies during your journey."

Fenella's stomach plummeted like a stone pitched into the loch. *Encounter Talfryn and the pixies while I ride with Mr. Blackwood? Oh nay!*

Blackwood, his expression tight, rode up to them, Fenella's sable mare at his side. Derry edged the fairy horse closer.

"I fear there will be nay more time for flowers on this journey," Fenella told Vendel as she mounted.

"They grow south of Urquhart Castle," Derry said. "We will ride right past them."

Fenella skewered the maid with a fierce look.

Derry, seemingly oblivious, dismounted, then slipped a ribbon from her hair. "Let me aid you, my lady." Derry tied the tansies to Fenella's saddle. Then, with the flowers tended to, Derry returned to the fairy horse and swung up onto the tapestry blanket.

When Fenella cast a final glance at the dwarf, Vendel's sunken eyes brightened. "Along the Loch Ness. Aye, I'm sure ye will see them there, milady."

Nay! I wish nay more fairy encounters!

Men and horses grew restless behind her. "Thaness." Blackwood sounded irritated. "The day grows long. We must be on our way."

Vendel bowed, "Mab speed."

Derry's eyes widened, and she pressed her palm over her mouth.

Fenella's brows twitched at her maid's odd behavior. "Godspeed, Vendel."

The contingent plunged ahead. Fenella lost sight of the dwarf as, once more, guards surrounded her.

"Mab?" Blackwood asked as he drew his horse alongside Fenella's.

Fenella shrugged, as if she had never heard of the Queen of Fairy.

While Blackwood urged his steed to the front, Fenella strained to see the path ahead. Worry stiffened her shoulders. *If we encounter Talfryn, how will I explain him to Blackwood?*

The contingent rode through Inverness, then followed the sweeping river south to Loch Ness. Trapped within the phalanx of guards, Fenella had little hope of spotting Talfryn. She watched for a gap between the riders, or a chance to ease to one side.

"We are nearing the Red Bard's land."

Fenella started at Blackwood's voice. He slowed his horse and signaled for her to join him. The guards parted, allowing her horse to trot to Blackwood's side.

"Keep your wits aboot you. Red Bard is friend to neither the Blackwoods nor the Roses," he warned in a low voice.

Remembering her mother's tales of how the Roses had plundered the lands around Urquhart Castle, Fenella nodded. Unease ensnared her chest as she pictured her mother in happier times. She glanced heavenward at the darkening sky. "The horses grow weary. Will we pass through this land afore the sun sets?"

"Once we are well south of the castle, we will rest for the night." Blackwood stood in his stirrups as if to stretch his legs and better search the terrain. "We will ride along the ridge," he continued, dropping back down into his saddle, "and keep the trees between us and Red Bard's stronghold."

'Tis a solid plan. Pines stood shoulder to shoulder down the steep slope almost to the water's edge. If the contingent kept to the ridge, she might ride past Talfryn and never know it. *Good… Nay. A glimpse wouldna hurt, just to be sure he is well again.* She twisted the reins.

The *whip, whip, whip* of wings sounded overhead. A raven dove for the flowers tied to her saddle, stole one, and flew into the woods with it.

"How peculiar," Derry said.

"True—"

Without warning, the fairy horse spooked, then bolted down the heavily wooded slope.

"My lady!" Derry screamed.

"Derry!" Fenella's muscles tightened. She reined her mare towards the descent, forcing the guards around her to halt or move out of the way. Derry's screams carried up the steep slope.

"She will alert the Red Bard's guards!" Edward kneed his horse. Fenella followed him down through the woods, leaning back in the saddle to counter the steep decline. The company pressed in around her. Angus and a few other guards rode ahead, threading through the trees at breakneck speed, either to silence Derry or protect Blackwood.

Derry's screams stopped.

A breeze rustled the trees. The tangy scent of pine engulfed Fenella. Nay glimpse of Talfryn's spiky yellow hair. Nay pixie cackles.

Fenella pressed her heels against the mare's sides, urging it to hasten. The jangle of bridles warned her the guards had also quickened their pace.

If I do nae distance myself from the Blackwoods and their Cold Iron, she silently scolded, Talfryn will never approach.

Racing full tilt, Fenella's mount sped downhill and breached the tree line. Her horse's hooves left the forest's soft pine duff and crunched hard onto the narrow pebble beach. The sudden view of vast blue-gray water shining with boundless beauty awakened Fenella's blood and stole the breath from her lungs. She slowed her horse, and as she stared at the sheer immenseness, the still waters of Loch Ness stirred.

Chapter Fourteen
The Creature in the Loch

Spanning the width of the bay, a great, slow-churning wave rolled in and submersed the shoreline. As the deluge flattened, smaller waves trickled onto the shore in its wake, slurping and plopping against half-submerged rocks afore receding. 'Twas a strange tidal pattern, Fenella thought, since nay wind came off the loch.

"Fate's wings." Fenella spied Red Bard's castle on the headlands up the coast. If she could see *it*, then the castle guards could see *her*. She drew within the tree line where she eyed the sun. Would night's cloak come soon enough?

Glancing back over her shoulder, she spied Derry astride the bright fairy horse deep in the woods, surrounded by Blackwoods and their dark steeds. Tension eased from her body. Somehow, she had ridden past them. Mayhap the Blackwoods had hung back because of the monster rumored to live within these waters.

Fenella scanned the loch's surface. A large, dark log bobbed across the orange trail of sunlight glittering on the water. Mid loch, a trio of ducks paddled away from the log, headed for the shore. Her mare whinnied and raised its head. Fenella's heart cantered as she looked up and down the coast, searching for any signs of danger.

When the ducks reached the shore, they took a few hurried steps, then flew into the woods.

Her palfrey backed up a step, blowing air through its nose.

I must find a glen or stream—some place amiable to fairies. With fixity of purpose, Fenella whispered to her mount. "Away with us." But the horse stiffened and would nae budge. Perplexed, Fenella glanced uneasily at the stronghold across the inlet, heedless of the tumult transpiring in the woods behind her.

"Stay within the trees!" Blackwood approached at a swift speed and rode up beside her. "Make haste!" Startled, she flinched as he grabbed one of her reins and kneed his mount. Guards pressed in on all sides, forcing her deeper into the tree shadows. The sharp odor of sweat fouled the air.

One of the lead horses shied, crashing against another steed and rider.

"Mother of—"

Edward released his grip on Fenella's rein, and called to Angus, "Control your mount!"

The horses neighed. Some bolted out of the way, others bucked. Derry shrieked. Fenella clutched her reins and edged her mare out of the melee. Angus's steed flattened its ears and kicked a nearby horse.

"Angus!" Blackwood's voice sounded low and lethal.

"The animal has gone daft!" Angus lost his stirrup, and afore he could slip his foot back into the iron triangle, the black bucked.

"Bluidy—" Angus landed in a bed of pine needles. His mount galloped into the trees amid a loud rumbling. Staggering to his feet, Angus lost his balance and fell again. "The ground is trembling!"

When the earth near Angus tore apart, Fenella gaped at the raw gash. Around her, more horses bolted. Men shouted. Her mare reared.

"My lady, beware!" Derry cried.

With a loud crack, a diseased pine teetered towards Fenella. She braced for the worst, but the sable palfrey lunged through the tree line and raced towards the water. The angry ground pitched again. Her mount faltered, nearly throwing her.

"Thaness!"

Holding tight to the reins, Fenella leaned low over the mare's mane. A short distance away, Derry clung like a barnacle to the fairy horse.

Ahead, the water swelled and flung towering waves against the shore. Fenella stared past the breakers to the middle of the loch where one end of the log rose out of the roiling water, arched its serpentine neck, then turned in her direction. Through wet blinking eyes, the creature seemed to study her.

Fenella's heart thundered to a near stop. She could nae feel her legs. The world grew small, as though she were peering through a keyhole, and eerily silent, save for her horse's shallow breaths.

Mayhap satisfied, the monster broke eye contact and dove beneath the water. Its slick black arched spine surfaced in the fading light, then disappeared again. Foam crests shot up on either side of its submerged body, marking the creature's swift path through the loch.

"'Tis headed our way!" Fenella yelled.

The earth gave another hard shake.

Too late, Fenella smelled her mare's fear and felt its muscles bunch. The palfrey reared and launched her into the air. Landing on one foot, Fenella tumbled sideways. With a crack of pain, air whooshed from her chest when her back slammed the ground.

Nearby, with a frantic flick of its tail, the fairy horse shied, nearly unseating Derry afore it retreated, carrying the maid to safety. Fenella forced a breath into her lungs and, in a flurry of linen and wool, rolled away from her sable's deadly hooves as it bolted for the woods.

"Fenella!" Fear and concern sharpened Blackwood's voice. He reined his horse towards her.

She rose to her knees. Fast-moving crests of water neared the shore. Once more, the monster's dark head surfaced. But this time nigh the beach, looming close to where Fenella struggled to remain upright.

"Edward!" Fenella stumbled to her feet, bruised and breathless, and nearly tripping on the hem of her long cloak.

Now afore her, the monster raised its head high and arched towards her, sending bright blue water, a hue she had only seen when water fairies revealed themselves, streaming down its slender neck and spattering her cloak.

As the air thickened and warmed about her, Fenella breathed in the loch dragon's tang. *Hyacinths!*

The sun paused on the horizon.

Fenella squinted at the creature. Surely it could nae be a water fairy. 'Twas far too big. She recoiled, remembering the Lady of the Lake's handmaiden, and how Glynis's sea dragon arm mark had come to life and attacked her.

The horrific memory sent a violent shudder coursing down Fenella's body. Pebbles at her feet rattled. As she gaped, her shudder continued, upturning sand and rocks in a direct path to the water. When it reached the loch, the unbidden force pushed back an incoming wave.

Splashed by recoiling waves, the monster stared at Fenella in apparent surprise and...*disappointment?*

"Fenella!" Blackwood's shout returned her to the present.

Edward dismounted, grabbed her around the waist, then swept her up. His strength and protectiveness wrapped around her like an invisible shield, diminishing her terror 'til it lost its power over her. The world narrowed down to the thumping of his heart against her side as he carried her to the trees.

"Are you hurt?" Blackwood asked, his heart beating so quickly Fenella feared 'twould rupture.

"Nay." She marveled at how safe she felt in his arms. "And you?"

For a heartbeat he tightened his hold. "Unharmed."

"Thank the Fates." A relieved breath escaped her lips. She became aware of his steady gait. Blackwood's limp, she realized, had disappeared while he carried her, as if with her, he was stronger—whole.

He halted beneath a tall pine and lowered her to the ground so she could stand but kept his arm around her waist. Brow furrowed, Edward scanned the shore and woods afore returning his gaze to her. "Ride with me 'til we find your mare."

Fenella's heart flipped at the determined set of his jaw and the assumption of leadership in his stance. "I thank you."

With precision, the remaining Blackwood guards encircled them. Fenella noted a quarter of the men were missing. Derry squeezed in beside her afore Angus pushed through the circle, leading Blackwood's horse.

A mournful wail issued from the water. None of the others, save for Derry, appeared to hear the sound, and none glanced back at the loch, save the maid and Fenella. The thaness clutched her mantle when she imagined the mysterious monster threw her what seemed a sorrowful glance afore it slid below the surface. An iridescent blue sheen spread atop the waves, then dissipated. Fenella felt an odd pull towards the water as dizziness whirled her insides.

A raven flew low overhead. She looked up when the bird uttered a hoarse kraak and a knocking sound afore it disappeared above the trees.

"Fenella?"

She shook herself, then dropped her gaze.

As Edward cupped his hands to boost her atop his horse, a rude noise sounded high in the pine.

Pixies!

Chapter Fifteen
Old Friends

Fenella flung the hood of her cloak over her head and tugged it low across her eyes. Afore she could warn Blackwood, brightly colored pixie dust bounced off his head and splattered the ground.

"What in bluidy—"

"Oh my!" Derry exclaimed.

From beneath her covering, Fenella held her breath and watched, helpless, as Blackwood slumped to the ground in a deep sleep. *Thud. Thud. Thud.* Derry, the horses, and the men fell in rapid succession.

She exhaled the moment the dust settled, then pushed back her hood and peered up into the nearest pine. A pair of close-set eyes within a small, pale green face, shined triumphantly.

"Puffball! What have you done? Blackwood's horse could have crushed him when it fell." Her gaze swept the fallen. Derry appeared asleep, but uninjured. Only the fairy horse remained standing.

"Don't worry." Puffball waved her bony hand at the collapsed Blackwoods. "They'll wake up in an hour or so." The diminutive pixie chortled.

Squashy-faced Nettle elbowed Puffball aside. Grasping a cluster of pinecones for support, the pixie swept off his dappled hat. "Thaness."

"'Tis good to see you Nettle, but was this truly necessary?"

Both pixies bobbed their heads with vigor.

"I am sorry, Thaness." Vendel emerged from behind a tree. "Desperate times demand desperate actions."

Fenella crouched beside Edward and splayed her hand over his heart. Her stomach unclenched when his chest rose, then fell against her palm. Reassured, she stood. But a new dread replaced the old. *What does the dwarf mean, desperate?* "Where is Talfryn?"

"Down here."

Fenella's insides fluttered. She twisted towards the Halfling's voice. Downslope, a familiar, lithe figure appeared. Her breath hitched. The worry and guilt that had knotted her stomach for months vanished. She ran towards him.

Halting within arms' length of Talfryn, Fenella exclaimed in a hushed whisper, "I feared you were dead!"

"I tried to follow you to Avalon, but my godmother magicked me to Fairy so I could recover from the Cold Iron poison."

"Och. Of course." Fenella frowned at the memory. "Have you only just recovered? Is that why you waited so long to appear to me?"

Talfryn threw the others an anxious glance. "Has it been long?"

"A year!" she exclaimed.

He rubbed the back of his neck. "Time moves differently in Fairy. To me, it seems far less."

She had settled enough disputes among her clan to ken when someone was telling less than the whole story. Backing up a step, she studied him. *Distracted. Yet he appears hale.*

His daft yellow hair still stabbed the sky like moor grass reaching for the sun. Though his cheekbones were more prominent than she remembered, his face almost gaunt. And she had nae noticed afore the luster of his skin, nor the lushness of his lips. *He seems more… ethereal. More fay.*

But most striking was his gaze. Talfryn's upturned eyes gleamed unnaturally bright. New colors rimmed his gray irises—the soft violet of fairy orbs, the swirling greens of treetops seen in magicked flight, and the stark indigo of a moonless night.

Breathtaking.

A bewildering sense of rapture swept through Fenella. Then Gran's warnings about the allure of fairies sprung into her mind. Nigh on swooning, she tore her gaze from Talfryn's beguiling eyes, then strove to compose herself.

Talfryn flashed her a questioning look.

Is he oblivious of his enthralling effect?

After a moment, she ventured a half step closer. "You should nae be here."

"Because of him?" Talfryn glanced upslope and scowled at her betrothed.

"Because you are endangering us all." She seized his arm and pulled him towards the trees, away from the Blackwoods, but hidden from Red Bard's castle.

"Wait!" He halted, like a horse refusing to budge, forcing her to halt, as well. "My father is in great peril. I need your help."

"Beacon? What ill has befallen him?"

"My mother, and mayhap my father, stole a potion from the queen. Their misdeed occurred years ago, but Mab just discovered the theft. She placed a binding cuff on my father that robs him of his magic and his ability to leave Fairy."

"How dreadful!"

Nettle and Puffball scurried to Fenella's feet.

"Even worse," Talfryn continued, "Mab chained him to a rowan with Cold Iron. Every two days 'til we return, more Cold Iron will be added."

She pressed her palm to her chest. "But he will die."

"We have 'til the next full moon to find and return the potion," Puffball explained. "Five days!"

"Five days, Fairy time," Vendel corrected.

"The Cold Iron will probably kill Beacon afore that," Nettle fretted.

"And if he is still alive, and we don't find the potion," Talfryn said, "Mab will make an example of him. Throw him to the Unblessed Court."

"Imagine what the Unblessed will do to a noble from the Blessed Court!" Puffball wailed.

"Enslave him." Nettle shook his head.

"Our legends tell of an Unseelie Court," Fenella said, "malevolent fairies who assault humans. Is that your Unblessed Court?"

"Aye. Bogies. Boggarts. Kelpies. The ilk," Talfryn said.

"Why would Mab consort with such creatures?" Fenella asked.

Talfryn exchanged uncertain glances with the others.

Nettle lowered his voice. "No one knows for certain...or is willing to say."

"But..." Puffball interrupted in the same guarded tone. "The tribute began right after Mab married the last true Fairy King—"

"Who died soon after their wedding," Talfryn explained.

"On the same night the king's advisor, Gudbrand the Dwarf, was murdered," Vendel added, his tone dour.

"How suspicious," Fenella said. "Where is the potion now?"

"Queen Mab believes my mother has the vial."

"But we don't know where Disa is," Puffball said.

"Mab banished her," Fenella remembered. "Doona she know where she sent her?"

"Our queen sent her to a Norse village, but she doesn't know where, or in what century," Talfryn said.

"Because she neither cared nor expected to see Disa again." Vendel's tight expression revealed nay affection for the queen. "Had Mab known of the theft, she would have murdered Disa on the spot or enslaved her."

Talfryn swept back his cloak and pointed to his waist belt. "I brought the Finding Horn. It will show us where my mother is. Hopefully, you'll recognize the village."

"You must understand the Norse sailed to many lands." Fenella needed to sit down. "The only time I have been outside this kingdom was when I fell down the wormhole and landed in Fairy."

"But the Norse did settle here?"

She glanced at the loch. "Nae this verra spot, but in the islands to the north, and in the southern kingdom." She twisted her mantle in her hand. "Even if I recognized the place, we would still need to know which year. The Finding Horn canna help us with that."

Some of the hope died in Talfryn's eyes. "We have to try."

The sun, a glowing orange ball, touched the horizon. "'Twill be dark in a moment," Fenella warned.

"Then please make haste." Talfryn tugged on the silken cord at his waist and removed the sleek amber vessel. He ran his thumb over the intricate elven metalwork encircling the rim and pressed one of the decorative dots. The stopper released with a satisfying click.

Their eyes met.

"'Tis too dark," Fenella protested. "We willna be able to see anything."

"Just try." The emotional storm swirling in his eyes pulled her in. "Remember? Only one can look, or the image will vanish."

Fenella drew cool air into her lungs. "You look first."

"Nay. I won't recognize the surroundings. Besides," his voice dropped to a hoarse whisper, "if it shows her grave…"

Thoughts of her own mother's death rushed back to her. "Verra well." Fenella cupped the horn with her hands.

Talfryn turned away. The pixies scrambled up him like he was a young willow and settled on his shoulders. Vendel faced the loch, but nae afore Fenella noted the intense look in his eyes. She sensed the dwarf had his own reasons for finding the Norsewoman.

Fenella flipped back the stopper and peered into the dark watery depths of the Finding Horn. "Show me Talfryn's mother."

Chapter Sixteen
A Debt Owed

"Her name is Disa," Talfryn said.

"Verra well. Show me Disa, wife of Beacon, mother of Talfryn."

The liquid in the Finding Horn rippled. Then a large loch, or mayhap the sea, appeared. Torchlight glimmered on the inky waves. A dragon baring fangs inched into view; its nostrils flared. Fenella gasped.

Talfryn whirled and clasped his hands around hers. Precious drops of the magic liquid splashed his thumb.

"Sorry!"

Averting his gaze from the Horn, Talfryn asked, "What did you see?"

"A dragon!"

Talfryn's hands clenched. "And my mother?"

"I doona ken." She exhaled a calming breath. "Turn away, and I will try again."

With seeming reluctance, Talfryn released his grip and twisted so his back was to her.

"Are you certain it is a real dragon?" Puffball asked.

"Nay… I am nae sure." Steeling herself, she peered more closely. The dragon's eyes appeared hollow. Its sleek neck bore intricate carvings instead of scales. *By the Stone!* A hot wave of mortification washed over her. "'Tis the prow of a dragon ship."

"A Norse dragon ship?" Talfryn's voice cracked.

"Aye. Your mother must be on it."

"See land?" Vendel asked. "She may be on a large river."

Without warning, a monstrous wave lifted the tall prow heavenward, then disgorged it into the watery depths. "She is at sea!" More of the ship came into view. A woman, her windswept blond braid thick against the circular shield she wore strapped to her back, appeared near the prow. She held a raised torch in her hand. But the image vanished. Fenella released the tense breath she had been holding.

"Gone," Vendel surmised. His shoulders drooped.

"Aye." Fenella tipped the stopper into place, then listened for the lock to engage. "Your mother is alive somewhere in the distant past." She pressed the Horn into Talfryn's hands. "Find someone with the Sight to guide you to the right time and place."

"The Lady of the Lake?" Talfryn wondered aloud.

"Viviane resides in the distant past. 'Twould take too long to find her."

"Who then?" Talfryn asked.

Fenella fisted her cloak. "The Romani."

They all stared at her with blank, confused expressions.

"The Gypsies," she explained. "Their caravan will be headed for Fife or, mayhap, Galloway. They could be near or a considerable ride away. 'Tis difficult to ken." Fenella heard Derry moan and darted her gaze to the fallen Blackwoods. The maid rolled over on her side.

Talfryn lowered his voice. "We don't need steeds. Mab has restored my magic." He linked his arm though hers. Vendel reached up to clasp Talfryn's other arm. The two pixies scurried under Talfryn's cloak and dove into his tunic pocket. "Ready!"

"Nay!" Fenella broke away. "I canna accompany you."

As though taken aback, Talfryn stared. His gray eyes darkened like a gathering storm. The air sharpened and a sudden wind chopped the loch.

"I will nae break my vow nor endanger Mr. Blackwood again." She strode towards her betrothed.

"Endanger?" Talfryn stepped towards her, glanced at the Blackwoods' Cold Iron, then stepped back.

"Aye." She trembled, remembering the hurt and betrayal in Edward's eyes, the physical pain he had endured. "Great harm came to him the last time I disappeared."

Vendel approached Blackwood's prone form. "He appears hale to me."

Fenella crouched near Edward's head and smoothed back a lock of hair that had fallen across his face. "He suffered great pain and became lame because his father wouldna allow his leg to be treated. Had I nae returned when I did…"

"Who risked his life so you could return in time?" Talfryn asked.

"You did." Fenella rose.

"And my father guided you through the veil."

Heat flushed her skin. "It took me months to win back Mr. Blackwood's trust and affection!"

"Then, mayhap you never had it." Talfryn crossed his arms over his chest.

Fenella turned away, stung as if slapped. Memories of Edward…and of Talfryn…crashed through her mind. *I never break a pledge, and I once vowed to never bring harm to Talfryn's family. But the cost this time…* She blinked back tears.

"Take us to the Romani," Talfryn coaxed. "Once they have told us where to find my mother, I will return you, forthwith.

Blackwood will never know you were gone."

Fenella glanced at Edward and the others asleep on the ground. "I canna leave them like this. What if the loch monster returns?"

Talfryn snorted. "Agnes? She won't harm them."

"Agnes?"

"Yes, she is the serpent of these waters. All of Fairy knows this."

"But the Red Bard *will* harm them if he finds them on his land." Fenella stared at the castle bathed in flickering torchlight.

"If Talfryn wakes them, they will see ye whisked away by magic," Vendel pointed out.

Asleep, Edward and Derry will be at the Red Bard's mercy. Fenella wound the edge of her mantle around her scarred hand. *And the laird has nay mercy for Blackwoods. Awake...* She could nae bear to betray Edward.

Vendel scratched his chin beneath his short beard. "Can yer magic wake them slowly so they do not see us disappear?"

Talfryn shrugged. "Aye."

"Then do it. The thaness will know her betrothed will be safe, and we'll return her afore much harm can be done."

Fenella bit her lip. 'Twas risky. Blackwood might think she had betrayed him, or he might think the Red Bard had stolen her.

"Beacon will die without your help," Talfryn pressed her.

Anger blotched Fenella's cheeks. "Promise me they will fully wake within a few minutes. And Edward must waken afore his men," she insisted.

"Why?"

"Because those are my terms." *If he is the last to wake, he will appear weak to his men, and worse, to his father when the tale is told.* She raised her chin.

Talfryn glared. Fenella held his angry gaze 'til he flicked his hand. Yellow dust like a hundred miniature suns rippled over the horses, Derry, and the sleeping men. The maid and a few others stirred.

"And Mr. Blackwood." Fenella's voice was like Cold Iron.

With a rebellious set to his jaw, Talfryn hurled a yellow fist-sized ball of magic dust.

The spell struck with an *oomph* as if Blackwood had been punched. He rolled onto his side, clutching his stomach. "Fenella," he whispered.

Her eyes brimmed. *Brutally awakened, and I was his first thought.* Guilt and worry entwined her chest like twin serpents.

Vendel grabbed her arm and dragged her like a reluctant horse to Talfryn. "Knot yer mantle so it doesn't fly off," he warned.

Fenella wiped a tear with the edge of her plaid, then secured it round her waist.

"Close your eyes," Talfryn whispered while clasping her hand, "and think of the gypsies we seek."

Edward moaned on the ground behind her.

"The quicker we leave," Vendel urged, "the quicker we return."

Fighting tears, Fenella pictured the Romani who traveled through her thanedom. An odd sensation fluttered her wrists, and her hands grew light as though they might float away. She lost feeling beneath her knees. Then, surging magic catapulted her skyward.

Chapter Seventeen
The Romani

Fenella squinted against the howling gale while the dizzying bird's eye view whooshed her stomach and knocked aside worries over Blackwood. Her feet skimmed treetops. The haunting cry of an owl reached her ears. As she and the others whipped through the darkening sky far faster than a horse could gallop, Fenella's skin tingled.

Magic hurled them north. In minutes, they retraced the path Edward's contingent had taken. The loch. The forest. She choked on smoke wafting from cook fires. Frigid air set her teeth chattering "Slow down and li...li. . .listen for lutes and violins."

An ethereal shadow, Fenella searched the dark landscape for the Romani's campfires. The air warmed slightly. Music, haunting and foreign, floated up to her ears. "Here!"

Weight filled her body as a forest glade rushed up to meet her feet. She thudded to the earth, her soles stinging. Talfryn tightened his grip and kept her upright.

"Hide!" Vendel pulled them to the ground, his stubby hands strong and calloused. The three sank to their haunches behind a wild rose thicket. A frightened pheasant sounded a ringing *cork-cork* cry, then took flight.

Talfryn brushed a feather from his sleeve. "Have we found

your Romani?" he whispered.

"The music sounded right. Wait, I will see." Cautiously, she peered over the thicket. Long wooden wagons and bell-shaped tents formed a semicircle around a crackling cook fire. The lone violinist placed his instrument in the back of a wagon, then joined the other men sitting by the fire.

A girl dressed in a faded shirt and several layers of long, colorful skirts walked behind the men and ladled garlic-scented stew into their tin bowls. The girl's dark, elbow-length hair shielded her face from view.

Fenella shifted, snapping a sprig of heather beneath her boots. A one-eyed dog she had nae noticed afore stepped out of the shadows and barked. Two men rose to their feet.

The girl with the stew turned towards Fenella. Firelight glinted off her coin necklace and cast a warm glow on her dusky skin. With relief, Fenella recognized the lass's knowing eyes, her straight nose, and full mouth.

"Milena!"

One of the men took a step towards them. *"Gadje!"*

"Wait, Poppa!" Milena rested the cauldron on the ground. The dog growled. "Razvan, silence! Poppa, 'tis the Thaness of Thorburn. You remember. Sorley's cousin."

Fenella held her breath. Milena's father strode past the others and stared at Fenella in the moonlight.

"Stand tall, girl, so I can see you better. And the rest of you, show yourselves, as well."

Talfryn and Vendel rose beside her. Milena's father stroked his chin. "What have we here?"

"May I present Talfryn and Vendel." Fenella glanced to be sure the pixies were still hidden in Talfryn's pocket.

"Which one of you is Talfryn?"

Talfryn cleared his throat. "I, sir."

"We have come to see Stanka," Fenella quickly said.

"Ah!" The man's blackberry eyes glittered. "You wish to see the *drabardi* to have your fortunes told."

She nodded.

The Rom waved them forward and called over his shoulder, "Milena, tell your grandmother to ready herself for visitors."

Milena gave them a quick glance, then darted into a bender tent on the fringe of the semicircle. A moment later she reappeared and beckoned to them. "Come. Stanka is ready."

As she had seen Milena do, Fenella walked behind the seated men. Talfryn and Vendel started to follow, but Milena's father grasped their shoulders.

"Men do not walk behind," he said. "Go through."

Fenella ducked into the dimly lit tent and breathed in the mingled scents of spice, smoke, and freshly cut saplings. The young trees, many still bearing leaves, had been bent and tied to form the frame of the tent. Smoke escaped through a hole in the top.

At the far side of the fire, an old woman slowly rose from her low three-legged stool. Firelight glinted off the gold coins she wore around her neck. Her back hunched slightly, and her hair, once thick and black, now hung from her scalp in skinny gray braids.

"We meet again, Thaness." Her gnarled hands smoothed her shawl.

"'Tis an honor, Stanka. May I present Talfryn and Vendel?"

Flames from the fire danced in Stanka's dark eyes. She dismissed Vendel with a quick, curious glance. Then her gaze fell upon Talfryn like a gull's fixed on a herring. "Show me your palm," she ordered.

Talfryn bent low to squeeze through the tent opening. Far too tall, he remained hunched as he lifted his hand to the old woman.

Stanka drew in a sharp breath. "Come closer to the fire so I may see your future better."

His hand clutched in Stanka's, Talfryn bent over the fire. The *drabardi* motioned to Fenella. "Give me your palm, Thaness."

Startled, Fenella held out her right hand. Seeing the faint, key-shaped scar on her palm, she quickly switched hands.

Stanka raised one eyebrow. She gave Fenella a long look afore drawing her hand next to Talfryn's. The crone's thin lips compressed into a sour line.

Fenella's eyes met Talfryn's. Something jolted and rippled within her. She dropped her gaze.

Stanka's grip tightened, and her nostrils flared.

"What do you see?" Fenella asked.

"Three possible paths for you, Thaness."

"And for me?" Talfryn asked.

"Two paths." Stanka's gnarled fingers dug into Fenella's wrist. The old woman's gaze darted between the two palms. "Thaness, what is the name of your betrothed?"

Nae trusting Stanka's sudden change in tone, Fenella hesitated. Instinct warned her nay good would come of revealing Edward's name. She shook off Stanka's grip and backed away.

Talfryn, seeming to follow her lead, also pulled free.

Stanka tossed a yellowish-green spore into the fire, creating a bright flash. "His name," she demanded.

An odd sensation, like multiple fingers tapping on her head to shake loose the truth, gripped Fenella. Though fighting it, she revealed, "His name is...Edward."

Stanka hurled another spore into the fire. The sudden flash lit up the tent.

Vendel leaned into the tent. "His name is Edward Blackwood of Argonshire."

Chapter Eighteen
The Sight

Two paths? Talfryn could not imagine any path other than one leading to his mother and the release of his father. Beyond that... As a curl of smoke rose to the hole in the top of the tent, he shook his head to clear his thoughts.

Talfryn studied Stanka. *Why cause a flash fire?* Remembering how ceremonial magicians created distractions to disguise or launch their spell work, he glanced about the tent. The cramped space, and the bright colors and patterns in the rugs and pillows bore nay resemblance to how he imagined a magician's austere chamber should look. His suspicions rose when he detected the bitter taste of gentian in whiffs of the flash fire's smoke.

He followed Fenella's furious glare to Vendel, who stared blankly through glassy eyes.

"Vendel!" Fenella sounded angry and aghast.

The dwarf blinked. His flat features shifted from vacant to startled, then confused, as though he had just awakened and did not recognize his surroundings.

The crone placed a Truth Telling spell on him!

Stanka beckoned Vendel forward. "Your palm."

Talfryn stepped aside, his head bent to avoid colliding with the roof of the tent.

Vendel lumbered into the tent and appeared to gather his wits afore extending his hand.

"How unexpected," Stanka said after studying Vendel's palm.

"What?" Talfryn asked.

Fenella leaned in.

Stanka stared into Vendel's eyes, then appeared to choose her words with care. "One path—noble, yet dangerous." Her eyes, filled with the weight of foreknowledge, met Talfryn's afore she returned her attention to Vendel. "Beware. Your cause will cost you."

Vendel dismissed her warning with a scowl, then retreated to a spot on the far side of the fire.

Fenella rubbed her palm. "Stanka, we need you to use the Sight to find Talfryn's mother."

The *drabardi's* keen stare swiveled his way. "Human or fairy?"

Talfryn flinched. "Human." He pushed the Norse circlet down his sleeve and over his hand. "This was hers."

Stanka closed her eyes and weighed the armlet in her open palm. Her brow furrowed, plowing new wrinkles above her nose. "I see gold coins drop into a calloused hand, a Roman centurion's hand." Stanka cupped the armband. "The coins reach a monastery by the sea. A dragon ship slips ashore with the tide." The old woman winced. "Invaders swinging battleaxes storm the friary. They steal gold coins, silver candlesticks, and a goblet encrusted with rubies. I hear screams and smell fire."

Stanka's eyes wrenched open. "These are not good men."

"Mayhap," Talfryn said, his voice as dry as milled wheat. "But the coins were later melted down, and from their gold this band was forged. The woman who wore it is good."

The fire snapped, and somewhere in the darkness behind the tent, horses nickered softly.

Stanka turned to Vendel. "This good woman, it is she *you* seek?"

Vendel nodded.

"I seek her," Talfryn clarified. "Vendel and Fenella are aiding me in my quest."

"So it would appear." The old woman beckoned to her granddaughter.

Milena, who had lingered near the tent opening, hurried to her grandmother's side. Stanka gripped the girl's shoulder and sank wearily onto the wooden stool. After smoothing her skirts, Stanka shifted her gaze to Talfryn. "I can help you find your mother, but the cost will be high."

Of course, it will.

"The price is half of whatever you have in your pockets."

"Agreed."

"Nay!" Vendel and Fenella said in unison.

"Do not worry," the *drabardi* assured them. "If you return safely from your quest, I will return half of what I have taken."

"Agreed," Talfryn repeated.

"Good." Stanka handed him the armband. "Keep this ready. I will need it in a moment."

He slipped the gold circlet over his wrist.

"Milena, fetch another stool."

The girl left the tent, but quickly returned with a low three-legged stool. At Stanka's bidding, Milena placed it beside the fire, between Stanka and Talfryn.

"You have a pocket in your cloak?" Stanka asked.

"Aye," Talfryn answered.

"Place its contents on the stool; then I will proceed."

Talfryn withdrew Beacon's small silken pouch and placed it upon the stool. Vendel frowned.

Stanka sniffed the fay sack, then untied its silky threads. The shimmering blue-violet fabric opened like a bellflower on a summer morning, revealing a small mound of glittering crystalline specks.

That is not pixie dust, Talfryn decided, seeing the specks for the first time. *The color is wrong. Mayhap it is some sort of fairy dust?*

"Do you carry a dagger?" Stanka asked Talfryn.

He shook his head.

"I thought not. Thaness?"

Talfryn shrank back as Fenella handed over her father's jewel-hilted dirk.

Stanka pushed the blade through the sparkling mound. "Milena, my small tin platter."

Milena fetched the disk, polished it with the hem of her skirt, then handed it to her grandmother.

"Choose a half for yourself," Stanka told Talfryn, "and place my half on the platter."

Talfryn eyed the split mound. One pile was slightly larger than the other. He glanced up. Three sets of curious eyes watched him. He used his finger to knock some specks from the larger pile to the smaller, making them even. "You choose."

Stanka clucked her tongue in approval, then transferred half of the granules to the platter. Talfryn tied the ends of the pouch together. Once he had pocketed the shimmering sack, Stanka said, "Each of you. Pull a hair from your head and place it on the platter."

Fenella laid a strand of her long coppery-red hair next to Vendel's sandy strand. Talfryn winced as he plucked a bright yellow hair from his scalp, then placed it alongside Fenella's.

The old woman wove the three strands together, then tied the ends to form a circle. Next, she pinched a small amount of the crystalline specks. "Everyone, be seated around the fire."

Milena removed the extra stool.

Fenella sank to the top layer of rugs and crossed her legs beneath her long skirt. Uneasy, Talfryn crowded past Vendel so he could sit beside her.

Once they were settled, Stanka said, "You see together; you journey together. Watch the fire closely." With that, she hurled the glittering dust and circle of hair into the blaze's heart. The fire hissed and sparked. Blue-green flames curled around the twist of hair.

"The circlet," Stanka demanded.

When Talfryn gave her the gold cuff, Stanka cupped the armband tightly within her hands, then swayed as though in trance.

Aqua and bright blue vapors funneled out from the fire, then curled along the rugs and pillows. Salty sea scents permeated the air.

"'Tis like the Viking mist I saw in the Finding Horn," Fenella whispered.

Vapors swirled around them, cold and damp. Talfryn noticed Fenella shiver next to him, then the fire drew him back. He no longer heard the men outside murmuring over their stew or the horses neighing. The rippling of a sail, the creaking of wood, and the slap of waves assailed him.

"Find Talfryn's mother," Stanka whispered, her focus intent upon the wrist cuff.

In a burst of purple light, the Sight sprang from Stanka's palms, whirled thrice around the Norse armlet, then snaked into the fire's blue-green core.

Talfryn tilted towards the vision unfolding in the flames.

The Sight skimmed over a firth, reached land, then raced past a forbidding castle perched on a cliff. Talfryn grew dizzy as it snaked up and down hills, passed through a tremendous forest, then undulated over Hadrian's Wall.

Churches and castles blurred afore him. The Sight scattered sheep asleep in a pen. It slipped through fairy hollows and musty dragon dens, then snaked into a candlelit room.

Talfryn clasped Fenella's wrist. "There's my father." Near Beacon, a female dwarf handed a newborn babe to a woman with blond hair and eyes the same peculiar shade of gray as his own.

He leaned forward. "That must be—"

"Disa," Vendel confirmed, tilting towards the fire. His voice softened. "And Hilde."

"You knew my mother?"

Stanka hissed at them to be silent.

Beacon knelt beside the wooden birthing chair and dabbed his wife's reddened face with a fluffy, moist cloth.

"Name the babe afore Mab returns," Hilde warned. She covered the afterbirth, then hurriedly washed and dried her hands, using a pitcher on a low stand bearing a glowing opaque bowl and a storm-gray drying cloth.

"Talfryn is a strong name." Disa looked into Beacon's eyes as though seeking confirmation, and with a gentle nod he seemed agreed. Clad in a sweat-soaked linen chemise, she rocked the infant from side to side.

Rosemary and lavender scents wafted from the fire and filled the tent as Talfryn stared at the vision afore him.

"Is this yer cloak?" Hilde interrupted, pointing to a mantle thrown over a mushroom-shaped stool.

"Aye." Disa stilled. Her expression hardened. "I loosened the sewing and left you the needle. Is the seam large enough?"

Hilde withdrew a stoppered crystal vial from a cord hidden beneath her neckline and held it against the cloth. "It might fit. Barely." She bent over the cloak. Her thick braid of tawny, white, and a few black strands, fell over her shoulder. "In ye go," she coaxed the vial as she wedged it into the narrow opening. "Done!" She grasped the needle.

"Have care. Do not lose the rune, stone heart, and crystal sphere hidden in the pocket."

"I will take heed," Hilde assured her.

"Talfryn," Beacon whispered against his son's damp hair. "You need a secret fairy name, as well."

Hilde swore, then sucked on her finger. As the dwarf bent over her sewing once more, Beacon whispered into the baby's ear. The infant's tiny fist uncurled.

With a final tug, Hilde tied off the stitching. She handed the cloak to Beacon, warning, "Take care ye do not break the crystal vial."

In a blink, Beacon replaced Disa's soiled chemise with a conjured clean tunic, leggings, and after a pause, a padded, brown-orange overtunic the same hue as newborn fawns. Then he draped his wife's cloak over her shoulders and pinned it with the brooch he had given her the day she had told him he was to be a father.

"I will guard the vial with my life." Tears trickled down Disa's face as she kissed her son's forehead, then handed him to Beacon. "Thank the Goddess he looks more fairy than human. Mab may spare him."

Beacon placed the babe in a small flower-bedecked bower near the birthing chair, then dropped to his knees again beside Disa. "My love, it is not too late. We can find another way."

Disa threw the midwife a desperate glance.

"There is nay other way," the dwarf declared, her tone pitched to quell second thoughts. "Otherwise, we would never have— indirectly, of course— leaked word to the Court of yer whereabouts and that the birthing had begun."

"Hilde speaks the truth." Regret and resolve threaded Disa's voice. "Someday, if I survive what is to come, I vow I will find my way back to you." She removed a gleaming band from her arm. Appearing to summon her strength, she twisted and tucked it under the blanket at the baby's feet.

Talfryn swallowed hard, rubbing the band on his wrist. Fenella squeezed his hand. Throat tight, he watched the scene in the flames unfold.

Disa unpinned the intricate brooch from her cloak and handed it back to Beacon. "Aside from your love and our son, this is my favorite gift from you. Safeguard it for Talfryn. 'Tis how I will recognize him if—"

"—When," Beacon corrected.

"*When* I see him again."

Beacon placed the brooch beside a necklace on a small inlaid table.

Talfryn noticed Fenella touched her neck and realized she still wore the duplicate chain and charm hidden beneath her gown.

"The queen will arrive soon," Beacon warned. With a wave, he transformed the birthing chair into a narrow bed with resplendent bedding.

"I must shift," Hilde said. "Scream!"

Disa screamed as though still in childbirth. The sound covered the dwarf's pained yelp as she shape-shifted into a tawny-eyed wolf.

Talfryn and Fenella gasped.

Beacon threw back a small floral rug, and the wolf bolted into a hidden tunnel. Quickly, he wrapped thick drying cloths around his hands, then pulled a sheathed sword and round wooden shield from the tunnel. With a word, he sealed the opening. After replacing the rug, he threw his wife an anguished look.

"You know why I must do this." Disa's voice cracked.

"Aye, to avenge Gudbrand and the Last Fairy King." He handed her the sword.

"My initial quest." Disa belted the weapon around her waist, then concealed it beneath the bedding. "But now that I know Mab lures humans into slavery—"

"It wounds me that you and your sister were once slaves—"

"We *are* slaves. Thralls. *Ambátts*. And I have nay knowledge of whether Bree still lives or has died. Retrieving the potion for the dwarfs is my only chance to find her. Grambolyn promised—"

"My love—"

"Afore you met me, you cared little about those held in bondage in my world or yours."

"Humans enslaved you, not Mab." Beacon's tone grew defensive as he handed her the shield. "As for my world, only our queen keeps slaves. It is a tradition Mab brought with her from her former kingdom. We have nay taste for her ways."

"Yet none in Fairy stopped her."

"We were too shaken by the sudden deaths of our king and his trusted advisor."

"Three centuries ago!" Disa glanced at the rug covering the tunnel.

Beacon conjured extra bedding to conceal the circular shield. "Mab, as you well know, is secretive and punishing of those who question her rule." Beacon seemed pressed to convince his wife that he was not to blame for Mab's actions, nor was he in a position to prevent the queen from doing as she pleased.

"How long have the three children Mab currently keeps been enslaved?"

"In human years? I am not certain. Decades, mayhap."

"Do you know any of their names?"

Beacon's face reddened. "No one has called them by name in a very long time, my love. But I vow, I will discover their names."

Tears shined in Disa's eyes. "Thank you." She sniffed. "Without her precious potion, Mab cannot trap more children into eternal enslavement."

"We could empty the vial. Replace the contents with something harmless."

"Nay. The dwarfs' mage requires the full potion," Disa argued. "She said the vial itself may provide clues to who killed the Last Fairy King and Gudbrand. I must return with the potion and the vial. Those were the terms."

"We have nay guarantee Mab will banish you to your own time and people. The dwarfs' mage may be long dead."

"I accept that risk. I must." Disa clasped Beacon's hand between her palms. "Mark my vow. Once my promise to Grambolyn has been fulfilled and I have found my sister, I will make my way back to you and our son."

"Your world is too dangerous for a woman alone without protection or magic." Beacon's tone was pleading. "Even shield-maidens need time to recover from childbirth."

Disa sank back. Her mantle fell open. Beacon conjured a stronger, less eye-catching brooch to secure it.

She lowered her voice to an urgent whisper. "Not since the Last Fairy King have the dwarfs had a noble on their side. You are their eyes and ears at Court. Remember, if a fairy kills another fairy—"

"She, or he, will lose immortality," Beacon said.

"Aye. We both suspect Mab killed the king and Gudbrand, then drank some of the potion to regain her immortality. If I have the potion—"

"Then she will not be able to cover the crime if she kills again." Beacon rubbed his thumb across the thunderstone he wore around his neck. "Mab will begin to age, and all of Fairy will know she is a murderer."

Disa squeezed his hand. "I pray to the All-Mother our child will not suffer because of my actions."

"Your scheme will likely cause us all to be killed. At best, our babe will be at Mab's mercy."

"She has nay mercy." Certainty frosted Disa's voice.

"Then reconsider. Release me from my promise."

"Nay. Beacon, you agreed—"

"That was afore we fell in love. Afore..." He glanced down at the infant in the bower, then lowered his voice to a faint whisper. "Once Talfryn is grown, if Mab has not been stopped and my king avenged, then *I* will do what must be done."

"Be sure our son is safe first." She pulled Beacon close for a desperate kiss. "Remember," she whispered against his lips, "you promised not to search for me, lest you lead Mab to—"

Frankincense, bitter orange, and cinnamon scented the tent. The flames flickered. A towering fierce-faced fairy with skin like a bronze coin and glittering dragonfly wings appeared in Beacon's hut.

"Mab," Vendel whispered under his breath.

Disa clutched her cloak and cast a heart-breaking glance at Beacon, then the baby.

The queen followed Disa's gaze. "The Halfling survived."

"Aye, dear queen," Beacon said. "A healthy male."

"Then the deed is done." Mab flung her arms wide.

"Nay!" Beacon shouted.

In a storm of golden dust, Disa vanished. The baby wailed.

Talfryn drew back, sucking air between his teeth. Fenella wrapped his hand in hers and gently squeezed.

"Follow Disa," Stanka commanded.

The Sight plunged through the golden dust, then entered a tunnel of white light. Talfryn sensed his mother's presence ahead as if the Sight were a hound on the shield-maiden's trail. The tunnel went black, then white, time and again.

"Three." Vendel mumbled.

Was that the North Star?

"Four."

When the Sight burst out of the tunnel, Talfryn heard angry shouts over a hail of arrows. Beside him, Fenella ducked and shut her eyes. A man recited psalms over the furious cries. A heartbeat passed, then another. Talfryn heard the creak of a wooden keel, the slap of waves, and the ripple of a sail in the wind…then nothing.

The tent fell silent save for Stanka's rhythmic breathing and the occasional hiss of the fire. As the mist curled into the blaze, the flames absorbed the vapors.

Talfryn, his gaze wild and unfocused, plowed his fingers through his hair.

Vendel rose awkwardly to his feet. "'Twas a huge help. Thanks."

Fenella and Milena stood.

"Talfryn?" Fenella pulled him to his feet.

"We must take our leave," Vendel insisted. He herded Fenella and Talfryn towards the tent opening.

"Wait." Stanka leveled her shrewd gaze on them. "You forgot something."

Fenella glanced about the tent. "Och, my dirk! I thank you." She slipped her father's dagger into her boot and stepped towards the opening.

"Nay," Stanka said. "Talfryn has forgotten my payment."

"But I gave you half of the magic specks," Talfryn said.

"You agreed to give me half the contents of your pockets, aye?" Stanka folded her arms across her chest.

"Aye—"

Stanka stared at Talfryn's cloak. "You have another pocket."

Chapter Nineteen
Talfryn's Choice

Fenella glanced at the tent opening and judged their chances for escape. *None.*

Puffball popped her apple-green head out of Talfryn's tunic pocket and pushed back the Halfling's cape with her spindly arm. "Is she talking about us?"

"You are in his pocket." Stanka's set jaw and determined stare did nae waver.

Fenella clutched the edge of her plaid. "Stanka—"

Nettle popped up, glanced at the *drabardi*, then mimicked her by folding his arms across his chest. His long, pointed ears twitched. "I am not staying."

"Me either," Puffball said.

"A bargain is a bargain," Stanka said in a calm yet raised voice.

The dread constricting Fenella's throat worsened.

A shadow scuttled across the tent, then Milena's father peered in. "Is there a difficulty?"

"I will let you know, Stefan. Bring the blackthorn staff. I believe its true owner has been found." She angled her head towards Vendel.

"As you wish." Stefan's blackberry eyes rounded when he saw the pixies, but he backed away, then disappeared.

"Trust me, Stanka," Fenella said. "You doona want either of them."

"Hey!" Puffball protested.

"We're wasting time," Vendel said.

Talfryn cleared his throat. "I do not think of Puffball and Nettle as 'contents', but clearly you do. Therefore, I will honor our pact and leave one of them with you. When I return, I'll expect you to return to me half of what you have taken."

"How do you know I will not keep your pixie and return the unused magic dust?"

Talfryn crossed his arms over his chest and widened his stance. "You can't return some of the pixie or some of the magic specks. You must return all the specks, or the pixie. You've already used some of the granules, so you must return the pixie."

Stanka frowned.

"Besides," Talfryn added, "I know my pixies."

"Would either of you like to offer yourself up?" Fenella asked.

Both pixies shook their head. Then Nettle mewed piteously. "It'll be me. It is always me."

Talfryn gently pulled the pixie from his pocket and held him in his hand. "Nettle, Disa is on a boat, at night. You hate night crossings, plus you become seasick."

Nettle's green face paled. Fenella worried he would be ill, but he appeared to muster his strength. "Puffball gets seasick too."

"Not as badly as you do."

Fenella followed Talfryn's gaze as he pointedly surveyed the tent. Was he purposefully drawing Nettle's interest to the curiosities surrounding them? Every pillow and rug had been woven in bright sunrise and sunset hues. Gourds and lanterns hung from the sapling frame. And the assorted bottles and covered pots begged investigating.

When he turned back to Nettle, Talfryn smirked. "Besides, think of how much fun you'll have 'til we return."

Nettle's eyes brightened. He hopped onto the large pillow. "See you in less than five days."

"Good luck," Fenella told Stanka.

"Wait!" Milena gave Fenella a small gossamer pouch. "'Tis Romani magic," she whispered in Fenella's ear. "In thanks for your past kindness to my people. Wear it around the neck to reduce fever."

The bag was so light Fenella suspected 'twas empty. Still, she tucked it in the sheath next to her dirk. "I thank you, Milena."

"To seal our bargain." Stanka slid an onyx ring from her forefinger and slipped it onto Talfryn's finger. "No iron," she assured him. "'Tis for protection. Wear always."

Talfryn glanced down at the black stone. Fenella stood close enough to see his reflection stared back at him. "I thank you," he said.

Puffball waved farewell, looking a little sad and guilty. Nettle absently waved back.

"We will return soon," Fenella promised Nettle.

The pixie overturned a wooden cup, ignoring her.

Once outside, they discovered Stefan waiting for them. "For you." He pressed a black walking stick into Vendel's hands.

"May I see?" Fenella asked.

Vendel held the staff at arms' length, level with his chest.

Fenella bent and examined the rough, twisted wood. The sharp thorns had been rasped down to smooth nubs. "'Tis made from blackthorn."

"'Twill protect you from ill-wishers and the evil eye," Stefan said.

Vendel blinked at the unexpected gift. "I will use it well. I thank ye."

Stefan glanced uneasily at Talfryn's chest, as if he expected Puffball to reappear. "May good fortune and helpful magic befriend you on your quest."

"Thank you. Farewell." Talfryn clasped Stefan's forearm, then led Fenella and Vendel out of the camp.

Once they had walked beyond the curious stares of the Roma, Talfryn abruptly turned to face Vendel, halting him, and asked in a low voice, "Did you conspire with my mother to steal the Aeternitas Potion? Is that why you volunteered to find her?" One-handed, he shoved Vendel's shoulder, forcing the dwarf to step backward. "Because of you and Hilde, Mab stole my parents back to Fairy while my mother was giving birth to me! Where were they? How did the dwarfs know of their whereabouts?"

Vendel raised the blackthorn staff. "'Tis not what ye think. Hilde told me yer father and mother magicked to Disa's early childhood home, the isle of I-shona. Long after Disa's family and the isle were lost to her, the island became the burial place of kings from the lands of her father and mother. Yer parents thought they would be safe amongst the dead. Although they avoided the fairy mound on I-shona…" He paused, glanced about, then continued in a whisper, "Glimmer tracked yer parents there. She expected ye, the heir to the Noble House of Storms, to further her plans by becoming an equistragon commander—" He stiffened and glanced about as though searching for eavesdropping spies. "Enough said for now."

"I agree," Fenella said. "Take heart in the knowledge Disa is alive."

"She chose a *cause* over her husband and child. We meant nothing to her. I mean nothing." Fenella heard the ache in

Talfryn's voice. And she felt his unleashed resentment and unbearable sadness in her own throat as she watched him advance another menacing step closer towards the dwarf.

Vendel pressed the blackthorn staff against Talfryn's chest. "Ye meant everything to her. Disa had been searching for another way afore Mab stole yer parents and brought them back to Fairy."

Fenella pushed the staff away. "If you knew that, then you should nae have betrayed them. You dinna give Disa a chance to devise another plan." She sighed. "But for Beacon's sake, we must put all that aside, gather our wits, and continue the search." The air seemed to still in this moment of decision.

"Now we know which century Disa's living in." Puffball had climbed onto Talfryn's shoulder.

"We doona ken that," Fenella said, releasing her frustration on the pixie.

"Aye, we do." Talfryn seemed to have cordoned off his fury and forced himself to focus once more on the task at hand.

"The tunnel went black four times." Vendel's tone suggested only a lackwit would fail to miss the significance of this fact.

Men! "Meaning?" Fenella asked.

"One century for each time the Sight traveled through a black phase," Vendel explained.

"Which century do you live in?" Talfryn asked Fenella.

"'Tis the fifteenth hundred and sixth year of our Lor—"

"Fourteen hundred, thirteen hundred, twelve hundred, eleven hundred!" Puffball held up four bony fingers.

Talfryn nodded. "Someone must know where a Norse ship would be headed in the year eleven hundred and six."

"Give or take a decade," Vendel said.

"I am nae so certain. 'Tis a late date for the Vikings. Disa could be almost anywhere. My Uncle Connell would know. He is a scholar of Norse history. But he left the castle afore we could speak.

"Anyone else?" Vendel asked.

"We could ask Eilif the monk," Fenella said. "He knows the Old Norse tongue and Viking history. And he has a map of Norse settlements."

"Let us find him," Puffball said.

Fenella shook her head. "Too risky. Eilif is like Gran. He has strong opinions against magicians, enchanters, and fairy folk."

"Then I'll throw a glamour," Talfryn said.

Vendel snapped shut a small device he held in the palm of his hand, then slipped it beneath his padded tunic. "Time escapes us."

"Verra well. But you must return me to Mr. Blackwood the second we leave Eilif."

Talfryn linked his arm through hers, and clasped Vendel's forearm. "Where can we find this monk?"

"Wait!" Puffball removed her skullcap. She snapped her fingers inside the cap, and two small drawstring pouches appeared. Then, she waved her spindly hand over the minute sacks, and both grew to three times their original size. "Pixie dust and fairy dust," she explained, tucking the pouches inside Talfryn's cape. "In case you are tricked into bargaining away the contents of your pockets again."

"I wasn't tricked." Talfryn pressed his palm to his pocket. "I won't ask how you acquired the fairy dust."

"Then I won't have to lie." Puffball returned her cap to her head, then asked Fenella, "Where dwells the monk?"

"In the village near Castle Thorburn."

CHAPTER TWENTY
THE MONK

Outskirts of Castle Thorburn

An odd sensation fluttered inside Fenella's wrists, and she could nae feel her feet. Magic tingled through her, tickling her skin as it whisked her up into the night. A bracing wind numbed her cheeks and made her eyes water. After a few breathless moments, she tensed for impact as they landed on the outskirts of her village. Lantern light glowed in many windows, and hearty smells emanating from evening meals rumbled her stomach. She withdrew, breaking the circle.

"Your monk?" Talfryn stared at the shabby hut set apart from a cluster of long, low houses.

"Aye." Her gaze met his. The wildness of the night wind and the threat of an advancing storm gleamed in his gray fairy eyes. Wonder, then worry threaded through her. "Disguise yourself."

Talfryn flicked his hand in a circle, and his cloak deepened from a shimmering bluc violet to a coarse mossy green. His spiky yellow hair dulled to a dirty blond and flattened. He flipped his hood over his head and tugged it low, shadowing his tilted eyes. "Ready."

Fenella approached the ox hide door. "Eilif," she called softly, "Are you aboot?" Not sure he heard or perhaps felt wary to answer a voice in the night, she whispered, "'Tis your thaness."

"My thaness rides to Argonshire." The voice from within the hut sounded suspicious. "Who dares speak in her voice?"

"Eilif, 'tis truly I, returned in secret."

A sword blade emerged from the edge of the ox hide and drew the animal skin aside. Stooped in the doorway was a tall wisp of a man wearing a tattered léine and trews beneath his mud-brown mantle. The light of a peat fire glowed behind him. His shadowy gaze fell upon Fenella.

"Thaness!" he said in a loud whisper. "Where are the Blackwoods? Did you flee them?"

"We were separated." With a pang, she thought of Edward near Loch Ness, surely awake by now, bewildered and worried. "Eilif, we must make haste," she said. "My friends wish to speak with you aboot the Norse."

Eilif peered past her shoulder. "Come in from the cold."

"This is Vendel," Fenella said as the dwarf passed through the doorway. She forced an even tone. "And Talfryn."

As Talfryn ducked through the door, the light from the fire glinted off his Norse brooch, catching Eilif's wild gaze.

"Where did you acquire that brooch, lad?"

"It was a gift from my father to my mother."

"Talfryn's mother is long lost to him," Fenella said in a rush. "He thinks her people came to this country at the end of the eleventh century or, mayhap, the beginning of the twelfth. We seek to ken her family home."

"Hmm." Eilif placed his sword atop a battered table, then pointed to Talfryn's brooch. "May I take a closer look?"

After Talfryn removed the brooch and handed it to Eilif, the monk eyed it greedily.

"'Tis fine work! Obviously centuries old. Yet, the gold is like new. There's hardly a scratch. Astonishing, truly astonishing."

Talfryn turned so his hood concealed his face, but nae afore Eilif gave him an appraising look.

"The early twelfth century, you say?"

"It was the date I was given." Talfryn shrugged.

"Hmm."

"If you please." Talfryn faced the monk, keeping his chin down so his hood shadowed his face, and extended his hand to receive the heirloom.

Eilif held the brooch above the Halfling's palm. "'Tis a bit late for the Norse."

"Nevertheless, the map?" Fenella prompted.

Afore placing the brooch in Talfryn's hand, Eilif resisted for a second more. A sinking sensation tugged Fenella's insides when she realized how long, elegant, and fay Talfryn's fingers appeared compared to Eilif's. As though taking note as well, the monk paused afore he dragged his gaze to a roll of parchment on a nearby scarred table.

"Aye, the map." He gave Talfryn a searching stare that belied his light tone. "Have you met our king?"

Puzzled, Talfryn glanced up from pinning the brooch to his cloak. "Nay."

Did Eilif see Talfryn's storm-filled eyes? Fenella watched the monk's face, but he showed no change in expression as he turned his attention to the table once more.

Eilif picked up the rolled map and tapped it against his open hand. "Quite a man, King James. He was nay older than the thaness when he became our sovereign. To this day, he is burdened with guilt. Do you know why?"

Fenella spoke up. "Everyone knows King James feels responsible for his father's murder."

"Aye, but do you know what he wears to punish himself?"

Talfryn shook his head.

"He wears a coarse iron chain around his waist. Can you imagine?"

Sweat pearled on Talfryn's temple. "Aye," he said in a shaky voice. "I can imagine."

"King James's father died nearly two decades ago. I wonder how many iron links young James had to add each year as he grew into a man. All that *Cold Iron.*"

When Talfryn swayed, Fenella pushed a wooden stool his way. "Eilif," she snapped. "Show us the map."

The monk's pale blue eyes glinted. "Of course, Thaness. Let me bring that stool to the table, and we'll have a look." Eilif laid his sword across the top of the parchment.

Fenella held back a wry retort at the monk's overt wiles as Talfryn stepped away from the table, and Vendel scoffed under his breath.

Eilif took his time gathering items from a cupboard afore moving the stool. When he finally settled himself afore the table, he weighted the lower corners of the map with turnips he had fetched from the shelf. "These are the Norse settlements."

Though Fenella had seen the map afore, she marveled anew at its well-drawn kingdoms and foreign place names written in spidery script. She recognized Norway in the upper right corner. An ink line ran from Norway down to the Orkney Islands, and from there, northwest to Iceland and south to Scotland. More lines looped around the Hebrides, marked "Sudrøyene," then to points south.

Talfryn remained a step back, avoiding the sword, and cleared his throat. "There are no dates on this map."

Eilif tapped his temple. "They are up here. Eleven hundred, did you say?"

All three nodded.

"Hmm." Eilif pondered the map for a moment, then stabbed his finger at a river port in Northumbria.

"Are you certain?" Fenella frowned. "My uncle Connell said—"

"I know more aboot the Norse than any Highlander ever will! *That's* where your friend will find his mother's people." Eilif folded his arms across his skeletal chest.

Talfryn frowned. "It is in the south, not in the far reaches of the northern kingdom."

Fenella's eyes met Talfryn's. "Aye. Which is why—"

"You came to me for direction." Eilif sniffed.

The dwarf waddled forward and stood on his toes. "This town?" he asked, pointing a finger at the map.

Eilif grunted.

Vendel locked eyes with Talfryn. "Jorvik."

Chapter Twenty-One
Jorvik

Unease skittered down Fenella's spine as she led the others away from the monk's hut along a moonlit path half hidden by junipers.

"Once in Jorvik," Talfryn said in a hushed voice, "we'll see if my mother's boat has arrived."

"Her vessel will nae be there." Failure knotted Fenella's shoulders as she halted beside a wych elm. She had sensed the monk's penetrating gaze sharp upon her back as she had hastened from the cottage. When she had glanced back over her shoulder, she had seen Eilif duck behind the ox hide, but nae afore she had caught the fevered glint in his eyes. "Eilif saw through your guise. He knows you are a fairy."

"Half fairy," Talfryn corrected her.

"One drop of fay blood is enough. He's directing you to Jorvik because the city's Cold Iron will kill you."

"Beacon said Disa would be in the far north," Vendel reminded him. "The opposite direction of Jorvik."

"My skin is crawling." Talfryn rubbed the back of his neck. "Something is pulling me towards Jorvik."

"Did the monk hex him?" Vendel asked Fenella.

"Eilif would hex if he could, but I doubt he has the knowledge."

"I tell you..." Talfryn scrubbed his arms as though cold. "A bewitchment grips me."

Fenella exchanged a worried glance with Vendel. "Nay soul in this village is aboot to cast an enchantment," she soothed.

"Smell that?" Vendel asked.

Fenella sniffed. "Sweet briar? None grows at this end of our village. The breeze must have carried the scent." As soon as the words left her mouth, she realized there was nay wind.

"We should remove ourselves." Vendel narrowed his eyes and gripped the blackthorn as though it were a sword he was about to unsheathe. "Now."

"Talfryn, if you are being bewitched, then I beg you to fight the enchantment." Fenella clasped his arm. Magic sparked painfully against her hand. She let go, as though bitten.

Shaken for a moment, she refocused, then said, "Think of Beacon. If you journey to Jorvik, you will waste precious time and place yourself in needless danger." When Talfryn didn't respond, she made her decision and turned to the dwarf. "Vendel, is Cold Iron poisonous to you?"

"Nay, Thaness."

"Then I beseech you, keep my friends safe. Talfryn, please return me to Mr. Blackwood. I urge you to then return north and search for Connell. My end of the bargain has been kept. Now, 'tis your turn."

Rubbing his arm where she had touched him, Talfryn said, "As you wish." He crossed his arms over his chest. Moonlight glanced off Stanka's onyx ring. "We will manage without you."

Stung by his curt tone, Fenella said, "Do as you will. I wish you good fortune. Now, magic me back to Mr. Blackwood."

Talfryn glanced down at his chest. "Puffball?"

The diminutive green pixie poked out of his pocket. "Still here."

"Good."

Fenella clasped Vendel's hand, and Talfryn grasped her other wrist. She gritted her teeth against the painful magic sparking off Talfryn.

"To Mr. Blackwood's side!" Talfryn's tone seemed discordant.

Without warning, Fenella's feet flew out from under her, and her stomach tumbled as though she had been shoved backwards off the ramparts. A scream rose from her core but failed to escape her throat. She lost feeling in her hands and feet. *Something is terribly amiss!*

The hint of apple, the scents of elm and juniper, and the silence of the night, vanished. Church bells rang, then abruptly stopped. War cries died mid sound. Smoke from harvest fires reached her, then were snatched away.

Horror and betrayal stormed her thoughts. *He is stealing me!*

Again, the clanging of swords, grunts, moans, and terrifying battle sounds rushed her ears. Fenella tensed like a bowstring. Memories flooded back to the day the Blackwoods had stolen her. She could still hear the clash of their swords and the cries of the wounded. A shiver coursed across her shoulders as she recalled the fever and body-wracking chills that had followed. The fear. Nay, the fury. Her grip faltered. "Hold on," someone said. It must have been Vendel, but in her mind, she heard Edward Blackwood's voice.

Sensations of hurtling backwards continued, fueling her anger. *How dare he!* Finally, the cacophony in her head gave way to the rhythmic snap of Talfryn's mantle and her own. She opened her eyes.

Vendel threw her a worried glance. Talfryn stared into the distance, unblinking, his eyes unnaturally glassy.

New sounds wormed into Fenella's ears: the slap of water against wooden hulls, merchants calling to one another. A cow mooed, and a raven uttered a hoarse *kraaak*. Straining, Fenella heard the quiet whir of a cup maker's pole lathe afore it abruptly stopped.

The stench of cesspits and tanning leather assaulted her. Fenella's stomach roiled. Soon thereafter, she thudded onto the ground, broke the circle, turned, and heaved.

"Mab's tears." Talfryn held back her hair as she retched.

"Doona touch me!" Fenella pressed the back of her hand to her mouth.

A damp ground mist swirled about. Single-story wood houses with thatched roofs surrounded them. Fair-haired women and bearded men, all speaking Old Norse, bartered around stalls in front of the houses. Fenella noticed their odd clothing. Each woman wore a pleated linen chemise beneath blue or scarlet woolen panels secured across the shoulders by straps and brass brooches. The men, a few dressed in chain mail, favored scarlet, blue, or leaf-green tunics and brown trousers.

A small crowd pressed in around them. Speaking in hushed tones, the men and women nudged each other. Their astonished stares fastened on Vendel. A blond man of about thirty, who wore a large Thor's hammer necklace, exclaimed in Old Norse.

Fenella had learned enough of the language to recognize some of the words. The man said something about the god Odin. "Sunlight" and "stone" were part of the mix. She wished he would slow his speech so she could understand the rest.

"I doona think they have seen a dwarf afore." She glanced down at Vendel, whose stocky body was half hidden by the swirling mist. "Since sunlight is falling on your head, and you have nae turned to stone—"

"Why would I turn to stone?" Vendel asked.

Several people gasped and stepped back.

"According to Norse legend," Fenella explained, "dwarfs are supposed to turn to stone when sunlight falls upon them."

Vendel snorted.

Talfryn, appearing dazed, said nothing.

"Since you have nae transformed, mayhap they think you are one of the dwarfs Odin created to hold up the celestial vault."

"The sky?" Vendel narrowed his eyes and frowned at the crowd.

"Aye," Fenella confirmed.

One of the Norsemen jabbed his axe into the air and shouted.

Pox! "He saw us fall from the sky." Alarmed, she searched their surroundings for a means of escape. *None.*

Many in the crowd grew wild-eyed. Those who bore weapons jutted them into the air and shouted. Fenella wished the Thorburn guards or her uncles' men were here. Since they weren't, she sidled away from Talfryn, slid the dirk from her boot, and hid the cold, hard blade up her sleeve.

Chapter Twenty-Two
Torn

In the Woods near Red Bard's Castle

Fenella! Edward sensed her nearby. Something akin to a battering ram slammed his stomach and he doubled over, but a hostile force kept him pinned to the ground. He whispered her name… Dread trembled through him. Muffled voices spoke on the edge of his dream. *Fenella. Two men.* Unable to fully wake, he gritted his teeth as he strained yet failed to understand their words. Had the Red Bard discovered them? He willed his muscles to cord and break him free. *Bluidy—*

Fenella's reluctance and despair washed over him. He moaned, sensing her slip away. *Nae again!* Struggling against the malevolent force, he finally slashed though it and opened his eyes. *Tree canopy…Loch Ness…Pain. Almost nightfall. Fenella?* Bolting upright, his stomach whirled as his surroundings spun. He dug his fingers into the sandy soil and clutched a tree root 'til the dizziness subsided. Then he slowly looked around. His contingent—what was left of them—lay sprawled on the ground. Horses. Men. His insides tightened. *Nay Fenella.*

Derry propped up on one elbow and glanced about. Her wide-eyed gaze met his in mirrored confusion. "Thaness?" Derry called, her voice sharp with worry.

Edward staggered to his feet, one finger pressed to his lips. He pointed towards the Red Bard's castle. Derry nodded. Carefully, he threaded between the fallen to reach Derry and help the maid stand.

"I see nay blood," Derry whispered. "Was this the loch monster's doing?"

"I doona know." A twig snapped. Edward pushed Derry behind him, then unsheathed his sword as he twisted towards the sound.

"'Tis us!" Three of his men emerged from the tree shadows. "Our horses carried us deep into the woods during the earth tremors," said one, who led Fenella's sable mare.

"Did you see Lady Fenella?" Edward asked in an anxious whisper.

"Nay," the man answered. "Just the palfrey."

The guardsmen dismounted and rushed to their friends on the mossy ground. "What befell them?" one asked.

"A sleeping spell?" Derry wondered aloud.

As the fallen men and horses stirred, Edward counted. All were present except Fenella. *Bluidy—*

Derry fingered the ribbon reins on the white horse. "Do you smell that?" she asked the mare.

"Bluidy flowers," Angus said as he stumbled to his feet.

"Hyacinths." Derry spun on her heel and stared at the loch.

"Nae now." Edward had nay time nor temperament to ferret out the meaning behind the strange maid's odd comments.

His horse neighed and flicked its ears back and forth. To calm him, Edward stroked the black's neck, then helped it up. The final rays of the sun glowed brilliant pink in the horse's dark eye.

As his men wakened, the last sliver of sun had nearly vanished beyond the loch. The vast sky made its swift journey from vivid pink, interlaced with dark gray cloud streams, to deep dusky-rose, blending into darker hues of gloaming, 'til all was doused in the darkest shade of black.

Nay hand could he detect afore his face.

Edward's contingent rose, and quick as able, gathered around him.

"Light your lanterns." Edward commanded.

The men made noises of turning towards their horses. Dim silhouettes felt the air 'til their hands met their mounts, then fumbled in darkness to procure their lamps from their travel sacks. Sparks from flint striking stone flickered among the men 'til wicks caught, then lanterns roared to life beneath ancient pines, birches, and oaks.

"Spread out and work your way up through the trees. Make sure your lights shine away from the castle and keep your voices low. Find Lady Fenella." He pointed to a highpoint of the forest. "We will meet there, on the ridge."

"Aye," the men chorused in whispered tones.

"Derry," Edward said. "Stay close." Then he moved to retrieve his own lantern from his pack.

"Aye, my lord." Derry mounted the white mare, peered into the shadowy forest, and shivered. "You can be sure, I will."

****

First light arrived in a blaze of vibrant pink clouds.

Edward paced along the ridgeline where his men could nae see the worry in his eyes. When last night's search had nae yielded so much as a scarlet tatter, the men, exhausted from the long ride, had sought shelter among the trees, taking turns sleeping, eyeing the woods, and watching the Red Bard's castle. Edward had nae slept. What he had heard and sensed during his half-dream state kept thundering through his mind like an impending storm.

Fear knotted his bowels. He tried to draw a breath, but the weight of his failure pressed upon him. *If the Red Bard's men had stolen Fenella, wouldna she have screamed? Mayhap they had taken her at knifepoint. Or did her kinsmen come for her?* His stomach clenched. *She ran away afore.* He rubbed the back of his neck. *Nay. She loves me. Though the death of her mother... What if my father...? Nay. His lordship would wait 'til after the ink had dried on our marriage contract. Bluidy—* Edward raked his hair behind his ears. *Should I stay and widen the search? Return to Thorburn and muster more men? Or ride south and face my father?*

The sudden neighing of the horses and a burst of movement among his men alerted Edward to travelers on the road. His gaze swept the forested slope, his scattered contingent, then the tamped earthen road beyond them. Three tall, enclosed wagons approached. He strode back down to his men, reaching them afore the slow-moving wagons arrived.

Though the caravan appeared innocent enough, no weapons in sight, Edward signaled Derry to remain hidden behind a tree, then for his men to follow. Angus and a few others swung up onto their horses, then kept their hands on their sword hilts as they ambled their steeds behind him to the side of the road. Edward stood afore his men and eyed the approaching lead horse-drawn wagon. A leaf and fruit design had been carved into its bright yellow sides. The old woman seated beside the dusky-skinned wagon driver leaned close and whispered in the man's ear. Edward slid his hand towards his sword. The driver signaled to the wagons behind him, then reined in his horse. The caravan creaked to a stop.

"Good morn," the old woman said.

"Good morn." Edward glanced at the men driving the wagons. *Romani. Gypsies.*

"Have you stopped us to exact a toll?" the old woman asked.

"Nay." Edward paused as his horse was led to his side. He swung up onto the saddle. "You are free to pass."

"But you have a request for which you require an answer."

The statement struck him as odd, and Edward flicked a sideways glance at Angus afore turning his attention on the gypsy. "Have you seen a young woman with bright red hair?" His pause was met with silence, so he expounded, "We became separated during the earth tremors. I fear for her safety."

"Mayhap the loch monster ate her," Angus muttered.

Irritated at this interruption, Edward frowned and held up a hand to quiet his father's man.

"The loch creature did not snatch the thaness, milord." Wisps of gray hair had come loose from the old woman's many braids. She hunched forward on the wagon seat and stared up at him with keen, dark eyes.

Edward's heart skipped two beats.

"The *monster* did not take your betrothed," she said more firmly.

"Then have you seen Lady Fenella? Is she in your caravan?"

"Nay, the Thaness of Thorburn is not with us." The old woman fingered the coin necklace at her throat. "But you may search the wagons if it will ease your mind."

He nodded to three of his trusted men to commence a quick search.

"Did Urquhart, the Red Bard, steal her?" he asked.

Her gray crescent brow arched upwards.

"Best pull out your purse if you hope to gain more," Angus whispered, then huffed.

Edward withdrew a half groat and a farthing from his purse, then tossed them to the silent driver. The Rom caught them with mute ease. He showed the coins to the old woman. She nodded, and the Rom shoved the money into his pocket.

"The Red Bard did not take the thaness. She was approached by an old friend who was desperate for a favor."

"What friend?" Jealousy flared like a blacksmith's fire. "How do you know this?"

Derry cleared her throat.

The Roma's gaze slid to Edward's purse.

Fumbling, Edward withdrew three more coins and pitched them. A half groat tumbled to the step beside the driver's boot. The coin glinted in the waxing light, untouched. The driver's blackberry eyes flashed.

Edward nudged Angus.

"What?" Angus protested.

Pointedly, Edward angled his head towards the fallen coin.

Scarlet-faced, Angus swung off his horse, retrieved the silver, and handed it to the driver.

The old woman kept her jaws clamped 'til Angus returned to his saddle.

"My lord, the thaness is nae in the wagons," said one of the men charged with the search.

"Well?" Edward prompted.

The crone smoothed her layered skirts, then tensely rested her gnarled hands in her lap. "The sun has risen, and we must break our fast. Mayhap you know of a safe refuge south of the castle where we can eat and speak in private?"

Edward inhaled a long breath, then slowly released it.

"The Thaness of Thorburn is kind and just. Like you, our only concern is for her welfare," the old woman added.

Edward held her gaze and judged her earnest, then said, "We travel south." He signaled the men behind him along with the rest of his contingent who had gathered at the back with the maid. "Derry, ride in the center."

"Are you serious?" Angus asked in a low voice.

"Deadly."

Chapter Twenty-Three
The Crone

Jorvik

The deathly stench of Cold Iron coated Talfryn's nostrils and watered his eyes. He blinked to clear his vision. With his skin itching and his stomach whirling from the bewitchment, he recoiled from the legion of Vikings afore him. A few hefted their weapons from one hand to the other. Hesitant mumblings grew into fear-filled threats, then boisterous anger. Thrusting their spears and axes, the emboldened throng closed in around them. Talfryn braced for an assault.

Unexpectedly, the shouts diminished into grumbles 'til the villagers fell silent, then parted. Into the void shuffled a stooped crone. Her wrinkled, bronzed face suggested a life lived outdoors.

"Hmm." Fingering the dragonfly brooch pinned to her cloak, the old woman aimed her beady gaze his way. Talfryn's skin prickled as she assessed him. Nearing his side, she raised her arm. A strange current shuddered through the gathering, buzzing Talfryn's ears. When the humming stopped, his ears burned like the afterglow of a hearth fire.

"'Tis a sign from the gods!" the crone proclaimed.

Several men cheered and raised their arms.

"I can understand them," Talfryn whispered to Fenella.

"Aye," she murmured. "The witch cast a spell."

Nonbelievers in the crowd clutched their Cold Iron weapons and slowly advanced. Out of the corner of his eye, Talfryn saw Fenella grab a sword from a nearby vendor's table and hold it ready.

"Step back!" Vendel shouted, pointing the blackthorn staff at the oncoming men. His walking stick rippled like a snake. Men and women, dropping their axes and spears, retreated, clutching their ears as if some horrible sound had assaulted them. Vendel jabbed the staff once more for good measure, then lowered it and jutted his chin.

The crone arched one brow afore turning back to address the crowd. "I will accompany the strangers and ascertain the gods' wishes."

A man with a two-fisted grip on his axe protested, "But—"

Air around the witch wavered, and the green apple smell of sweet briar spread like wafted incense. The crone glared at the axe wielder, and the Viking lowered his weapon.

"Return to your homes." The old woman waved her gnarled hand. "Welcome to your mother's world," she muttered as the grumbling throng dispersed.

Talfryn's scalp tingled. "What did you say?"

Thick wavering air turned to a soft shimmering aura. The crone's eyes slanted upward and lightened from raven black to bright blue. Deep wrinkles creasing her forehead and framing her mouth vanished as her weathered skin grew sleek. Tawny, ropelike locks replaced gray wiry wisps.

Glimmer, Talfryn realized, *Mab's sister.* Nerves brittle, he felt for the thunderstone beneath his neckline, careful to avoid the sharp talon. A potent storm captured in quartz, the crystal throbbed against his fingertips through the coarse wool.

Mab's sister flicked her hand, now ageless and long fingered, and, in a voice smoothed from a hag's rasp to a velvety caress, bade, "Let's remove ourselves from this Cold Iron."

As though caught gaping, Fenella closed her mouth. She hefted the sword back onto the merchant's table, then slid her dirk into her boot.

Talfryn exchanged a quick glance with Vendel. By silent agreement, they fell into step behind the Royal Sister, keeping Fenella between them.

The farther they retreated from the market center of Jorvik and entered parts of the town where wooden structures and mudbrick streets were more prevalent than Cold Iron, the easier Talfryn breathed.

While Glimmer strode through the dissipating ground fog, her wool Norse clothing transformed into violet gossamer wisps that concealed little of her body, bestowing upon her form a lightness, her posture lofty, her gait more lithe. Strings of bright gold teardrop markings emerged on her skin, encircling her arms, bare shoulders, and feet. Sparkling, delicate dragonfly wings, resembling lapis, purple, and teal stained glass edged in gold, sprouted from her back.

"Such a chill in the air." As soon as the words left Glimmer's mouth, a lush cloak of overlapping leaves, appeared on her shoulders. The remarkable garment unfurled around Glimmer's wings, molding to her body as it unfolded past her feet in a thick train.

Fenella clamped her hand over her mouth, apparently to smother a gasp.

They entered a deserted alley, and the shape-changing fairy turned to face them. "You are too conspicuous in those clothes,"

Glimmer said, referring to their highland garb, and seemingly oblivious to her own glittering appearance. "We'll start with you, Thaness."

"How do you know who I am?" Fenella asked.

Talfryn noticed Fenella was staring at the two rows of gold teardrops that dotted the fairy's skin from her hairline to the center of her brows, then continued like tears beneath her eyes. He touched her arm. "This is Glimmer, Mab's sister, and Godmother to all Trooping Fairies. She's one of The Six. The Royal Sister sees and knows everything." *Whatever you do, don't cross her.*

Glimmer's smile reached her high cheekbones. "You remember me! Astonishing, since I am so rarely invited to Court." She batted her eyelashes and dropped her voice to a falsely gay whisper. "Must not upstage our dear queen, nor bother her with details. 'Never undermine my authority!' So haughty."

Unsure of the proper response, Talfryn remained silent and merely blinked.

"My, my..." Glimmer fingered a lock of Fenella's red hair. A quizzical expression passed over her face, then she trembled like a bird ruffling its feathers. "Her hair, and her magic."

"Whose hair and magic?" Fenella asked.

"A sorceress from afore your time." Glimmer recovered her composure and flicked her wrist. Sweet briar scented the air. Sparkling burnt orange dust whirled around Fenella, then abruptly vanished along with the apple flower scent.

Talfryn released a long breath.

First, a pale green, sleeveless, pleated woolen chemise replaced Fenella's saffron tunic. Then, over the chemise, two panels of forest-green wool were joined across her shoulders with circular bronze brooches. Finally, her scarlet plaid vanished, and a fern-green shawl appeared in her hand.

"You look like a Norse free woman." Talfryn circled her. "Though I'm not sure the blond hair suits you."

Fenella reached behind her neck and pulled her windblown braid into view. "Fate's wings!"

"Don't worry," he said. "It is a glamour. Your hair is still red. The illusion will vanish as soon as you wish."

"Now for you." Glimmer faced Talfryn.

Remembering stories he'd heard about Glimmer's fiddling, he asked, "Godmother, may I manifest my own glamour?"

For a moment, Glimmer frowned and appeared offended. Talfryn's pulse skittered. But the Royal Sister leaned close enough for him to smell her gingery breath, then winked. "Proceed."

Puffball scrambled from her hiding place within Talfryn's pocket and leapt onto Vendel's shoulder. The dwarf squirmed as Puffball wormed beneath his shoulder-length hair to hide.

Talfryn's thoughts jumbled. A quick glance at Fenella in her Viking garb jarred him back. He raised his right hand above his head and drew a circle in the air with his finger. The fresh scent of juniper filled the alley.

The odd prickle returned, stronger, crawling his flesh, shivering his body. A sweet smell of green apples engulfed him. He threw Glimmer a quick sideways glance. Her fine features settled into an expression of extreme innocence. His stomach churned with suspicion and dread. All at once, pain knifed his arms, shoulders, chest, and legs. *She has melded her magic with mine.*

Glimmer's sapphire eyes gleamed. Queasy, Talfryn swallowed hard.

"Oh my," Fenella said. "You look so different."

Perplexed, he stared down on her. She had shrunk...or had he grown? Alarmed, he smoothed his hands over his new Norse garments. The itchy, woad-dyed woolen tunic and sea-blue breeches were warmer than the highland garb, though not nearly as fine as his normal clothes. At least he had retained his fairy cloak, now glamoured moss green.

Puffball pushed aside Vendel's hair and peeked out. Her eyes widened, increasing the upset in Talfryn's stomach. Spotting Glimmer, Puffball gulped, then shrank out of sight.

Talfryn tried to push his mother's gold armband higher up his forearm, but muscles rippling beneath his skin blocked the way. *Muscles?* Unnerved, he pushed his fingers through his hair. Instead of short spikes, his hair fell in a smooth wave to his shoulders. He pulled a lock within eyesight. Blond!

"An excellent glamour." Glimmer appeared a little too satisfied.

"Most convincing I've ever seen," Vendel proclaimed. "Ye might not be killed."

"About that," Glimmer said. "The thaness's monk was addled, else he deliberately sent you astray. I suspect a past unfortunate encounter with the Unblessed Court altered him. Nevertheless, your mother is not here."

"Do you know where she is?" Talfryn asked.

Glimmer shook her head, swaying her ropelike locks. "Somewhere among her own kind, I imagine." Her seeming indifference changed to scrutiny as she focused her attention on Talfryn. "The question is, what to do about you?"

"Me?" His pulse hastened.

"You…and your father."

"What about us?"

"The quandary is, I like you both. The Court *needs* you both. But should you succeed in returning the potion to my sister, there will be nay limits to her rule."

"Meaning her ability to eternally enslave humans?" Talfryn asked, remembering Mab's threat to enslave his mother, as well as the three human children with fading spell rings.

"That, too." Glimmer twirled a lock of her hair around her finger. "Yet should you fail, Beacon, and perhaps you, will die. Then the Noble House of Storms, our weather fays and equi-stragon commanders, will cease to exist. And we cannot have that, can we?"

Is that why you convinced Mab to steal my parents? You wanted me to be born in Fairy, under your control? He fisted his hands to halt the blue sparks racing down his arms.

"The potion has been missing for over eighteen years," Fenella challenged. "Has its absence diminished Mab's power?"

"Not in discernible ways—*yet*. But time is measured differently in Fairy than in your world, Thaness. To us, eighteen and a half years amounts to hours, mayhap days."

"If Mab is dethroned, who will succeed her?" Fenella asked.

"Brindleburr?" Talfryn wondered aloud.

Glimmer's amiable expression vanished. "Brindleburr is neither bound by marriage to Mab, nor is he a blood relative."

"So, *you* would ascend to the throne," Fenella stated. "Unless another member of The Six is a closer blood relative?"

"Fenella." Talfryn placed a warning hand on her wrist.

"No one is closer." Glimmer huffed.

"Aye," Vendel said. "So, absent a living relative of the Last Fairy King or Queen Titania…"

"The Hawthorn line is dead." Glimmer's cold dismissal sounded final. "If such kin existed—child or sibling—we would have…heard of them by now."

"We're wasting time." Talfryn extended his hand to Fenella. "I will return you to your betrothed."

"Not so fast," Glimmer commanded. "First, you must pass a test. Then, and only then, will I allow you to continue on your quest."

Talfryn's stomach dove towards his feet. "What test?"

"'Tis time you learned about your mother's people. You wanted to search for her? Then you must walk to the quay."

"Jorvik is filled with Cold Iron," Fenella protested. "You canna send Talfryn into the city. Besides, you yourself said his mother isn't—"

The stench of hostile magic thickened the air. A sudden gust scattered animal bones and other debris across the alley. Talfryn pushed Fenella behind him.

"We're sure ye will protect Talfryn, dear Royal Sister," Vendel said hastily.

The ill wind stopped as suddenly as it had started, and the cloying air returned to its fragrant scent.

Glimmer smiled sweetly. "Of course, I will." She slid a coiled serpent bracelet from her wrist. "There is no true protection from Cold Iron, but this will mitigate its affect for a short time. Just return afore the eyes glow scarlet."

With a faint hiss, the bracelet coiled itself around Talfryn's wrist, settled into position, then froze. The tail began to glow red.

Glimmer gave him a long look. "You may choose one ally to accompany you."

Talfryn stroked the mustache that had appeared above his upper lip. "I choose Fenella. She knows the most about the Norse and can blend in better than the other two."

"Figured as much," Vendel said. From beneath the dwarf's hair, Puffball whined.

Paying them no mind, Talfryn untied the Finding Horn and handed it to Fenella. "Keep it safe in case we become separated or anything happens to me."

Frowning, Fenella reached for the magic horn.

Glimmer flicked her hand, and a slender cowhide belt appeared at Fenella's waist. A clasp, made of the same unusual metal as the Finding Horn lid, dangled from the leather. "Nay time to dally," Glimmer scolded. "Use it to fasten the Horn."

The latch fit perfectly through a spiral flourish on the ornate metal lid.

"Take care of Puffball while I'm gone," Talfryn told Vendel.

Vendel thumped his blackthorn rod on the ground as though he were a palace guard called to the ready.

Talfryn squared his shoulders, which now felt heavy and broad. He strode ahead of Fenella, his stomach knotted, uncomfortably aware his steps were no longer airy or carefree.

Chapter Twenty-Four
The Second Ring

South of Red Bard's Castle

First light's rose-streaked sky had by now grayed into mid-morning clouds. *Fitting*, Edward thought as he glanced up, his mood equally dark. His impatience neared exploding. He needed answers, but the old woman would nae say a word 'til she had assembled her bluidy tent. The longer she stalled, the more his leg throbbed and his doubt flourished. For all he knew, she had been sent to delay him so the men who had stolen Fenella could ride farther away.

Except, while in his half-dream state, he had heard nay horses.

Ignoring curious stares from his men, Edward strode to the Romani encampment. With annoyance heating his blood, he barely noticed the cold breeze stirring the trees. He glanced at his horse tethered near the river with the other Blackwood steeds and Romani horses. The black swished its tail and lowered its head to drink as he passed.

When Edward approached the Romani, cook fire smoke carrying the scent of frying fish burned a path up his nose. He veered to avoid the vapors.

The wagon driver glanced up from the finnocks sizzling over the flames. After nodding in mute greeting, the Rom called out, "Stanka," adding more words in his foreign tongue.

Stanka peered out from the bell-shaped tent and gestured for Edward to join her.

Finally. Edward stepped to the shelter, then peered inside.

Freshly cut saplings draped in canvas arched over Stanka's head as she settled heavily onto a low stool. A small fire crackled afore her. Smoke wafted up through the hole in the top of the tent to escape, but a lingering spicy smell made Edward's eyes water.

"Sit and show me your palm," Stanka said.

Edward entered the tent, his body folded to avoid scraping his head. Once inside, he straightened as best he could but kept his distance. "I have nay time for fortune telling."

"Thorburn's thaness told me you were an honorable man. I wish to see for myself."

"She voiced those exact words?" Edward narrowed his eyes. "When did she speak with you?"

"Last night." Stanka waved him closer.

"You saw Lady Fenella *after* she disappeared?"

The *drabardi* met his glare. "Aye."

Barely controlling his outrage at her keeping this from him, Edward settled onto the second stool, his back rigid, his shoulders tense. Fuming, he offered up his hand.

Stanka studied his palm. "You love her."

Edward swallowed. "If you know anything, tell me."

She leveled her gaze at him. "The truth will not be to your liking."

He jerked his hand away. "Who stole the Thaness of Thorburn?"

"A rival beyond your reach." She paused. "But I know how to find him."

"Tell me."

Stanka leaned closer. "Your love—your faith—will be tested."

"Nay test can be worse than how you presently try my patience."

The gypsy arched one brow. "Your father controls Lady Fenella's thanedom, true?"

"'Til she marries or comes of age." He shifted on the stool.

"His lordship's men prevented us from crossing Thorburn land, a right that has always been granted us by the thaness as well as her mother and father afore her."

Edward leaned forward, one elbow on his knee. "I am certain Lady Fenella will honor your past agreement once I find her and we are wed."

The old woman straightened her spine. "You will not interfere?"

"Nay." Edward sat upright and expanded his shoulders. "You have my word."

Stanka nodded. "Lady Fenella vouched for you, so I will trust your word." She reached beneath a pillow on the floor beside her stool and extracted a small linen pouch, then cupped it in her hand like a bright yellow flower. "'Twas clear last eve that Lady Fenella's single goal is to return to you as quickly as possible."

Relief washed over him, and his heart quickened.

"Have faith in your betrothed, and she will find you, or die in her efforts."

Die? Edward's insides tightened. "Where is she? Who is this rival you claim has stolen her?"

The crone glanced down at the pouch in her lap, then leveled her gaze at him. Her dark eyes blazed with flames reflected from the fire. "The rival is unknown to you. You cannot touch him or find the thaness because he used…unordinary means to whisk her away."

"Unordinary? Do you mean magic?"

Without answering, Stanka inclined her head. "Your palm shows two possible destinies. First path: return home and trust in the thaness. Have faith she will find you. Or…" She stared into his eyes and dangled the pouch in front of him. "Second path: employ magic of your own and risk changing both of your destinies forever."

Edward frowned at the yellow pouch. *Do nothing or employ magic. Neither option is in my blood.* "How do I know I can trust you?"

Stanka shrugged. "Fate rules our destinies. She holds us accountable for our discretions."

"Fate has nay rule over me. 'Tis God I answer to."

"Take the pouch, and we will *have* to trust each other."

Eager to find his betrothed and vanquish his rival, Edward seized the bright yellow bag and slipped it unopened into a hidden pocket in his long shirt. "We shall see whose trust is well-placed."

Chapter Twenty-Five
The Norse

Jorvik

Clouds crowded the sun, casting shadows on the warren of wooden buildings that comprised the center of town. An overwhelming stench of open cesspits made Talfryn long for the loamy scent of the forest.

Rats scurrying towards foul-smelling midden kept to the edge of streets that rang with the Cold Iron shoes of workhorses striking cobblestones and the creak of carts hauling goods from warehouses along the quay.

Peddlers in open stalls shouted, holding up furs, silk, or amber, while artisans hawked intricate jewelry and wooden bowls.

Talfryn winced at the pinging and clattering of silver and gold coins that bore images of a sword, a bow and arrow, Odin's raven, or Thor's hammer when merchants dropped them into metal balancing pans to be weighed.

How do humans stand the din? At least his newly long hair covered his ears, dulling the noise.

Fenella nudged him. "We must cross the street."

Remembering his purpose for entering Jorvik, he refocused and followed her gaze to a large metalsmith's shop, ahead on the right. Men hunched over Cold Iron horseshoes, swords, crossbows, axes, and knives. Talfryn used his sleeve to wipe away the sweat beading his temple. The glint of Stanka's onyx ring caught his eye as he lowered his arm.

"This is madness." Fenella cupped his elbow. "We must turn back."

"Nay. I can manage." Glimmer's magic would have to hold. He didn't want to collapse in front of Fenella and his mother's people. Talfryn glanced down at his wrist. The bottom coil of the serpent bracelet glowed red, warming his skin. One and a half more coils afore the glow would reach the serpent's head. *There's still time to find her.*

Women thronged the streets in nearly equal number as men. And despite Glimmer's insistence that Disa was not in Jorvik, Talfryn searched each woman's face for his mother's gray eyes.

An old worry rose to the surface. What if his mother had never searched for him because she knew he would perish in her world? He clasped Fenella's shoulder with fresh resolve. "Tell me all the ways in which the Norse use Cold Iron."

"Beware of crowds." Fenella angled her head towards a group of men throwing animal bone dice. "The Norse carry iron weapons, nae always visible. Also, many wear buckles, clasps, amulets, or pendants made of iron. These items are more common among those of lesser means because iron is less costly than steel, silver, or gold."

A horse drawn cart rattled towards them on the street.

"You already know aboot iron bridle fittings and horseshoes," Fenella said. "And iron on cart wheels."

Talfryn's throat grew dry. "Aye." His voice cracked.

Fenella's brow creased, and she changed the subject. "We should find the quay."

"Aye." He quickened his step, fearful time would run out. Dodging and weaving, he kept as much distance as possible from each passerby. *The glamour must be working.* Only dogs roaming the streets gave him a second look.

"I am sorry for stealing you against your will." He lowered his gaze to his feet.

"It appears you were bewitched. Glimmer..."

He leaned and whispered in her ear. "In some ways, she is more dangerous than Mab."

"I sensed as much." Fenella's expression appeared more vigilant than furious, or so he hoped.

As they traveled west, his soft leather shoes, damp from the dissipating ground fog, offered little protection from the cobblestones, and he longed for the clean feel of moss and the morning dew on his bare feet. His thoughts drifted to the forest where his father had taught him magic beneath prickly pines, green beech, and ancient oaks. *I must not fail him.*

The street spilled onto the quay where a flurry of waterbirds squawked and whistled as, in seeming disarray, they dove past ships, dashed to shore, then darted back to sea. Wooden trading ships heavy with full cargos plied the waterway.

So many ships to search.

"Oh my! 'Tis grander than any vessels Connell described."

Talfryn followed Fenella's gaze to an elaborately decorated longship bearing a bronze weathervane.

"Must belong to a nobleman. And those," Fenella indicated three ships docked at the quay, "are *Karvs.* Uncle Connell said they were once used for raiding. By now, they are probably used for trade."

Thinking of Disa, he asked, "Nay dragon ships?"

"Nay. The *Drakkars* are war ships."

War ships? "Do you think my mother was taken prisoner?"

"I hope nae." Fenella's gaze flitted among the passing boats. "We must keep our distance." Her tone was once more wary. "Norse ships, too, are filled with Cold Iron. Sea chests are iron-bound. Mariners, whether traders or raiders, would have weapons forged from iron; even shields hewn from wood have iron nails. And anchors—"

"Enough!" Gulls circling overhead seemed to be laughing at him. Covering his ears, he fought the urge to throw stones at them.

"I am sorry," Fenella said, taking his hand in hers. "We have walked to the quay as Glimmer ordered. You have seen your mother's world, both land and sea, as she wished." Her gentle touch and soft-spoken words calmed him, even as he saw the red glow at his wrist. "Now let us hasten back."

"Not yet." he said. *I have time.* Quickly turning his attention to study the varied ships roped to the dock, then two new ones that sailed into view, one carved figurehead caught his eye.

"What about that one?" Talfryn released her hand and pointed to a high-sided vessel. "Is it a *Drakkar?*"

"Nay, 'tis too small."

"It has a dragon."

"Mayhap 'tis a chieftain's ship. I believe they were allowed dragon carvings."

He broke into a trot. "My mother may be on board."

Swifter than the flick of Mab's fingers, Fenella caught his arm and flipped back his cloak, exposing the magic bracelet. He saw what she saw–the red glow had nearly reached the serpent's head. Using a tone that told him she would brook nay argument, Fenella said, "We must return. Now!"

"But I can't leave without knowing."

"Your mother is nae here! Glimmer said so. Cold Iron—"

Talfryn clasped her hand. "The bracelet and Stanka's ring will protect me," he pleaded, then pulled her towards the boat as it docked, sidestepping a rattling cart loaded with coal.

When they drew near enough to see, Fenella spoke hastily. "The men are unloading cargo."

One Norsemen on the vessel handed down large wooden chests to men on the quay. A dock man stumbled. The chest he had been carrying fell and splintered open, revealing fist-sized chunks of raw amber.

"See? 'Tis a trader. I count five men on deck." Fenella persisted. "There would be nay room for more. Disa is nae there."

"Must be sure." Suddenly dizzy, Talfryn pressed his hand to his forehead and shut his eyes.

"Pox!" Fenella slipped her arm around his waist. "I knew we should nae linger."

He offered little resistance as she guided him back to the cobblestone streets. The creak of moored ships and the slap of water against oak hulls faded. A curly-tailed dog followed them, barking, 'til they turned a corner and headed towards the heart of the town where streets narrowed again and grew more crowded.

The stench of Cold Iron settled into Talfryn's lungs and throat, choking the flow of air. Bitter with disappointment, he slowed to a halt.

Fenella tightened her grip and pulled. Talfryn stumbled forward a couple of steps afore planting himself and refusing to budge.

"Cold Iron poison is addling your mind." She yanked harder and half dragged him past the metalsmith's shop. "I will nae leave you here to die." Talfryn felt blood drain from his face. He squeezed his eyes shut, desperate to keep them from rolling back inside his head.

"One foot in front of the other," Fenella urged, "or you and Beacon will perish."

Shouts from peddlers and artisans pierced the roar in his ears. Talfryn opened his eyes. *Almost there.* His skin was cold and clammy, except for the flesh beneath the bracelet, which burned. He dared not glance down at the glowing coils.

Breathing heavily beside him, Fenella alternately tugged and pushed his bulky Norse frame. The alley beckoned ahead. *Think of Father.* Talfryn forced air into his lungs and urged his legs to keep moving. *If I can just reach Glimmer…*

Blinding scarlet light from the serpent bracelet flooded the shadowy alley as they entered. Vendel and Puffball rushed towards them.

"Titania's tears!" Vendel exclaimed as he helped Fenella ease Talfryn to the ground.

"Speak not the name of the Last Fairy King's first wife!" Glimmer whipped her wand, causing sparks to snap the air near Vendel's head.

Winged blessings to the memory of our queen, king, and the Royal Newborn, Talfryn thought.

"You should nae have sent him into the city." The dangerous reproach in Fenella's voice made Talfryn wince. "He could have died from the Cold Iron."

"Precisely." Glimmer's heartless response stunned him. "Talfryn must understand the dangers of his mother's world."

Her sharp tone cut like an axe blade. "'Til he does, this illusion of what he thinks *may* be possible will keep him stuck between Disa's world and Fairy. As Supreme Weather Councilor and Talfryn's godmother, I needed assurance he will not work against his own kind."

"The fairies," Fenella clarified.

"Exactly." Glimmer fluttered her dragonfly wings. "It appears he has learned the lesson. His fantasy has ended."

Talfryn shut his eyes against Glimmer's stern words. Whatever small hope he had harbored about his mother's world had been crushed. His chest ached. Disa's world and Fenella's were much the same. He couldn't survive either.

Fenella's soothing touch trailed across his cheek. Her skirt rustled. Too tired to open his eyes, he felt her rise to her feet.

"You achieved your aim." Desperation tinged Fenella's voice. "Now help him!"

"I do not take orders from humans!"

Hearing the harsh rebuke in his godmother's voice, Talfryn half opened his eyes and saw Glimmer's hardened expression. A bright red beam sprang forth from the serpent's eyes, intensifying the ominous scarlet glow over them. Talfryn's arm burned as though a fire had ignited his skin. He groaned.

"Forgive the thaness, dear Royal Sister." Vendel shot Fenella a withering glance. "Humans can be overwrought with empathy. She forgot her place."

Fenella lowered her chin. "My apologies."

Glimmer, appearing appeased, flicked her elegant hand, and the fiery scarlet light retracted into the flaming eyes of the serpent bracelet. The jeweled orbs dulled into unpolished rubies, the adder coils cooled and grayed. Weak light from the somber sky again illuminated the alley.

Talfryn rubbed his wrist.

"Better?" Fenella asked, dropping back into a crouch beside him.

He managed a feeble nod. Puffball leapt onto his chest and mewled against his shoulder.

Fenella patted his arm, then rose to face Glimmer. "Supreme Weather Councilor sounds like a vital role equal, I dare say, to queen."

Glimmer's expression brightened and she appeared to relish Fenella's newfound deference towards her. "Quite equal to the queen, though Mab fails to realize it." Her expression grew cunning, as though she enjoyed a private jest. "My dear sister abhors gales, you see. She believes," Glimmer snickered afore continuing, "that I will *prevent* major winds." Glimmer slid a calculating glance at Talfryn. "Imagine..." She drew the word out for apparent dramatic effect. "If weather could be used to...say... summon unnatural mists to change the course of a battle. Or suppose a fairy could hurtle her enemies across oceans, continents, and even centuries."

His godmother's words wove in and out of Talfryn's ears. Surely, he hadn't heard her correctly.

"Sounds like something wizards or druids might conjure," Fenella argued.

"Wizards! Druids! Hah! They are ignorant of foreign magic. And they know nothing of equistragons, which are also under my purview." Glimmer huffed a dismissive breath out her nose.

"Equistragons?" Fenella asked.

"Fantastical beasts–part horse, part winged dragon. They are quite impressive with their tall, twisting horns. Their mere breath can cause scorching windstorms while their hooves and talons can kick up unimaginable whirlwinds." Glimmer glanced at Talfryn. "An equistragon's power could be harnessed for the Blessed Court's benefit should the beast choose a Trooping Fairy as its magical partner."

"I was chosen by Luminous," Talfryn muttered. With discomfort, he noticed Glimmer staring at the small bulge beneath his neckline created by the thunderstone.

"She's a winged horse," Puffball explained. "Born from the feather that fell from Pegasus when the great winged horse transformed into a collection of stars."

"Luminous cannot create whirlwinds and firestorms like equistragons." Glimmer's dismissive tone carried an unspoken warning.

"Luminous has other talents," Talfryn said, his voice pitched just above a whisper. *And I have the ability to create a whirlwind on my own.*

Fenella leveled a long, curious look his way.

Glimmer has cast weather magic as an almost military force. Father taught me equistragons and weather magic are meant to protect Fairy, not to wield cataclysmic weather against our enemies.

Fenella cleared her throat. "Your Royal Highness, Vendel and I can find Talfryn's mother for him. Cold Iron does nae harm us."

"Disa will only give the Aeternitas Potion to her son," Glimmer argued.

"I will find her," Talfryn mumbled. "You must return to Blackwood," he told Fenella. "Forgive me for bringing you to Jorvik instead of returning you to him."

The air between them quivered. Talfryn felt his true form, spiky haired and fay, shine through. Then pain tightened his throat. Death's icy fingers reached for him.

Fenella squeezed his hand, and the air settled as Talfryn's Norse glamour reclaimed him.

"Time runs short," Glimmer reminded them.

Talfryn struggled to his feet, waving away Fenella and Vendel's aid. "We'll return to Argonshire and find your uncle," he told Fenella. "If Mr. Blackwood is not there, I will magic you to him."

"I thank you."

Puffball clung to Talfryn's cloak. Vendel tucked the blackthorn staff into his waist belt, then took his place beside Talfryn. Fenella completed the circle.

"Wait!" Glimmer touched the adder bracelet. It rattled, and dozens of pinpricks pierced his skin. Talfryn clenched his jaw. "The armlet's magic will last two more times." Glimmer's gaze swept over them. "You'll not make it far in Talfryn's weakened state. And though I do deplore ignoring my dear sister's wishes to not meddle..."

Talfryn surmised the glint in Glimmer's eyes suggested quite the opposite.

"Even our dear Queen should not object to this." His Fairy Godmother crinkled her perfectly shaped nose and slowly raised her arm. "Especially, if no one *tells* her."

Vendel clasped Fenella's hand. Puffball closed her eyes.

Briar blossom perfume permeated the air as Glimmer made three circles with her briarwood wand, conjuring hundreds of white butterflies.

Boosted by this flurry of fluttering wings, Talfryn clutched Fenella's other hand, then commanded, "Argonshire."

Chapter Twenty-Six
Trooping Fairies

Magic yanked them aloft hard and fast, as if trying to gain enough momentum to sustain their flight. Almost immediately, Glimmer's butterflies fell away like snowflakes swept up in a gust. Hurtling through centuries as the bracing wind cut through her clothes, Fenella felt like catapult fodder. City stench gave way to loamy river smells, and the distant sound of Norsemen clanging metal turned to the roar of water tumbling into limestone gorges. Crisp forest scents tickled her nose.

To her relief, Talfryn lost his sickly pallor.

She heard the thud of stones being laid in abbeys being built. Women sang, their angelic voices rising beyond high stone walls and reaching skyward.

As they passed over the southern uplands, the air grew colder. The chop of carpenters' adzes and the ring of a blacksmith's hammer reached her ears. Castles and towns appeared. Seasons changed from summer to fall to winter to spring, the cycle repeating again and again.

Church bells pealed, then suddenly stilled.

A ripple of worry accosted her. Whatever boost Glimmer had given them was slowly losing strength. Talfryn and Vendel exchanged troubled looks. *So, they feel it too.* A weight that had nae been there afore caused Fenella to tighten her grasp on their hands.

They dipped low enough to hear grouse hidden in the heather call out. Then city smells, the sting of hearth smoke, and the acrid scent of hides being tanned assaulted her nose. She smelled corn drying in stone kilns, and the hawking cries of vendors on Market Day reached her ears.

Unseen gulls laughed, and the air grew clammy. Fenella searched the mist beneath her feet for blue-gray lochs.

The drag holding them back grew stronger, and Fenella again tightened her grip.

"Yer Norse glamour is weighing us down," Vendel shouted at Talfryn. "Change back!"

Talfryn compressed his face but gasped. "I can't!" He dropped lower in the air, pulling the others down with him.

"The wind is dying." Fenella sensed the water drawing closer. "We need a boost to blow us ashore."

"Trooping Fairies!" Vendel and Puffball said as one.

"I don't know how to summon from great distances," Talfryn said.

"Try! Maybe one is nearby," Vendel urged.

Talfryn puckered his lips and whistled a complicated series of bird trills.

Pressed against Fenella's hand, Vendel's palm grew slick with sweat despite the cool air and overcast sky. The four continued to sink into the mist.

"Wish I knew how to swim," Vendel pined.

Fenella steeled herself for the icy plunge.

"Look!" Puffball shouted.

Air snapped Fenella's cloak. She forced her chin down, gritting her teeth against the sudden gale that threatened to push her head back too far and snap her neck. Wind watered her eyes, and she was about to squeeze them shut when the flapping of enormous wings sounded over the tempest. Two tall, twisted horns pierced the clouds.

Fenella cowered, tightening her hold on the others as she drew back. A prickly puff of magic swept over her, sweeping through the others, too, judging from their expressions. Their descent now slowed. They seemed to hover atop or within an air pillow buffeted by strange supportive winds.

"'Tis an equistragon," Vendel shouted over the gusts stirred by the animal's leathery wings.

Fenella gaped. The creature's two rear legs ended in scaled hooves, and its two front legs—nay, *arms*—ended in terrifying talons.

The beast lowered its horned horse head as it pushed into the gale. What appeared at first glance to be translucent armor on the horse-dragon's head, throat, and flanks, proved to be scales reflecting the sky and clouds. Tiny sea glass bells, looped around the animal's neck, tinkled like wind chimes.

"Is that a Trooping Fairy?" Fenella asked, spotting a lithe figure with long, lavender tresses astride the huge horse-dragon. When she heard no answer, she glanced at her companions.

But Vendel seemed focused on Talfryn, who appeared too absorbed in the rider to hear Fenella's question. "Aye," Vendel said afore clamping his jaw shut.

Fenella squinted against sunrays glinting off the equistragon's horse mane and tail, which shimmered like liquid silver. The glare and distance prevented her from distinguishing the fairy's facial features. She wondered at the fay's tattered and blackened wings. "Has she been struck by lightning?"

Once more turning to Talfryn for an answer, but receiving

none, Fenella felt a clutch of pressure in her chest at the way he gazed upon the fay. Disbelief, pride, and gratitude seemed to play across his face like passing clouds. Curious of their connection, she drew her attention back to the figure riding the fantastical beast.

With tremendous might, the young fairy pulled back on the creature's mane and spurred the equistragon to claw the air with its deadly talons, fueling and stirring the windstorm 'til the coiling turbulence mustered enough force to funnel down from the sky. The whirlwind cut across the loch, leaving violent eddies in its wake. Herring, ripped from the water by the sudden whirlpools, spun in the air afore their frenzied wriggling bodies rained back down in raucous splashes.

As she watched the commotion churning towards them, Fenella's eyes stung from the hairs whipping across her face. Water sprayed her boots, and fish ejected from the loch thudded against her legs, knocking her ankles together.

"Hold fast!" Talfryn shouted, tearing his gaze from the fairy commanding the equistragon.

Crazed winds slammed into them, lifting them in a dizzying swirl. Teeth clenched, Fenella tightened her grip on Talfryn and Vendel. Hands locked, they spun around and around.

The windstorm tore Puffball's skullcap from her tiny green head and sent the hat careening over the loch. "I can't hang on much longer," Puffball shrieked.

Fenella fixed her gaze on the pixie and willed strength into Puffball's spindly arms. "Almost there! I see the stronghold!"

They whirled in rapid circles closer to the shore, wool rippling and flapping, water surging and splashing, fish flopping and thudding, and poor Puffball keening and mewing. As they spun low over the slim bank that divided the loch from the river, Fenella spotted a blurry row of trees looming below. Bile rose in her throat as dirt and rocks, kicked up by the wind, pinged against their trunks.

"Stop!" Talfryn shouted, his eyes locking onto the rider with each whirl.

The discordant roar ceased as the tempest halted. And their whirling slowed as they neared the trees beyond a low heath.

"My heartfelt thanks," Talfryn added in a tender tone that, to Fenella, seemed almost like a kiss.

The lavender-haired Trooping Fairy pressed her palm to her forehead, then extended her arm in Talfryn's direction.

"Hail and farewell!" Talfryn called out as the rider and her mount vanished. In their place, a sparkling, silver-rimmed cloud appeared.

Vendel released Fenella's hand and dropped hard to the ground near the base of a beech tree close to the tree line.

Puffball lost her grip, then flew like a small green arrow into a patch of Lady's Mantle. After retching, she staggered to retrieve her cap from atop a nearby rock, then half dove, half tumbled back into the clump of soft leaves.

Finally, Fenella and Talfryn landed roughly in a tangle. She released her grip on the glamoured half fairy as she fell back against heathery ground, pinned beneath his bulky Norse body.

"Sorry!" Talfryn said as he extricated himself. "I'm sure she tried to summon a gale, not a—"

"Aaugh." Fenella remained prone on the damp vegetation. Dazed and dizzy, she closed her eyes. Pain burned her shoulders and back.

"Fenella?"

The edge in Talfryn's voice spurred her to open her eyes. Fenella instantly regretted the move; the world spun around her, whooshing her stomach. She squeezed her eyes shut 'til the dizziness passed, then cautiously reopened them. *Better.* Air returned to her lungs. A glint of light blinded her, and she raised her hand to block it. Beyond her hand, she detected a long steel blade. She followed it to its tip pressed against Talfryn's shoulder. Her breath halted. Then she rolled heavily onto her side and curled her knees towards her chest to reach the knife in her boot.

"Niece, you are alive! Thank the saints."

She struggled to sit upright. "Uncle, remove your sword. Talfryn is my friend."

"A close friend," Talfryn added.

"Talfryn, you say?" Unexpected notes, a blend to Fenella's ears of incredibility and hope, tinged Connell's words.

Fenella nodded, causing a painful throbbing in her head.

Connell hesitated afore he withdrew his sword, then sheathed it.

Talfryn released a long breath. "My deepest gratitude," he said as he stood to help Fenella to her feet.

"The two of you—" Connell surveyed the herring and rocks strewn upon the green, registering surprise when he spotted Vendel making his way from the woods towards them. "The *three* of you best explain. To start, where is Edward Blackwood?"

"You have nae seen him?" Fenella cast an accusing glance at Talfryn.

His Norse clothing had vanished. In its place, he wore a Highlander's yellow shirt and a moss-green mantle. Hints of his Viking glamour remained in his long, wheat-blond hair and in the sinew that sculpted his normally lithe frame. He seemed taller beside her.

"Nay. I have just arrived." Remnant restless air rippled her uncle's saffron shirt. "I was aboot to ride up to the donjon to discuss your wedding with Lord Argonshire when I heard a commotion and veered this way."

Her shoulders, tight as a bowstring, eased as Fenella noted curiosity brightened Connell's eyes, nae anger, nor suspicion.

Cogitating how to respond, she twisted the hem of her scarlet plaid around her hand, numbly realizing she, too, was clothed again in highland garb. Only the cowhide belt remained to remind her of Jorvik.

"I became separated from Mr. Blackwood during an earth tremor near Red Bard's castle. We have been searching for him."

'Tis close enough to the truth.

"I see." Connell's gaze flicked from Fenella to Talfryn, then back to Fenella. Uncle Dunstan would have crossed his arms over his burly chest and intimidated her with a low growl. Sorley's father, Uncle Artair, would have nodded, then strode off. But Uncle Connell stood his ground and waited.

Pox. She freed her hand, then raised her mantle over her head, covering her bright red hair from prying eyes. They had landed beyond the riverbank, on a flat open area of heathers and grasses. A beech woods stood between them and the tall rise upon which the Blackwood stronghold sat.

Vendel waddled over to them and held out his stubby hand. "Vendel Fairheart, sir."

"Och. My manners have vanished. Vendel, meet my uncle, Connell Thorburn."

Connell bowed to clasp the dwarf's forearm.

Then Fenella glanced at the clump of Lady's Mantle. Puffball parted two stems and waved. Beyond her, fishing boats, still rocking from the whirlwind, dotted the loch, their nets cast for herring and their wicker creels waiting to be filled.

"Uncle, we need your help," Fenella said, as she brushed the dirt from her léine, then turned to him. "And I would rather nay others see us."

Fenella followed Connell's glance up at the Blackwood's stone donjon overlooking the loch. Only three arrow slit tower windows, one at each story, faced the woods.

"This way." Her uncle glanced back and made a clucking noise. His horse obediently followed as Connell stretched out his arms and herded them into the trees.

When nay person on the loch nor in the castle raised the alarm, Fenella exhaled a long breath, then filled her lungs once more. How much time had passed, she wondered, since they had left Edward and the others on the shore of Loch Ness?

"We need to ask you about the Norse," Talfryn said, addressing Connell.

Her uncle scrubbed his knuckles along the dark stubble on his jaw. "I will tell you what I can of the Norse." His gaze burned a path from the waning ripples in the loch to Talfryn. "But then *you* must tell *me* what *you* know aboot Trooping Fairies."

Chapter Twenty-Seven
Uncle Connell

"Trooping Fairies?" Talfryn's voice rose with forced bemusement as he sidestepped away from the horse and the Scot's Cold Iron. But his merry expression withered beneath Connell's unflinching gaze. Then Talfryn's thoughts jumped to his father, spellbound and at Mab's mercy. He hardened his resolve. "What do *you* know about Trooping Fairies?"

The Scot's muscular frame moved with rugged ease as he dropped into a crouch, the scabbard attached to his waist belt scraping the sparse wind-tossed grasses beneath the trees. "I had never heard of them 'til one year afore Fenella's birth." Connell cast his niece a fond glance, then plucked a stick from the ground as she sat down next to him.

"'Twas late one night I was on the road, and as I reached a fork in the path, lightning struck, startling my steed." He snapped the stick in two. "Caught unawares, I was thrown to the ground and hit my head. When I woke, I was shivering, soaked, and solitary. My horse had abandoned me." Connell shook his head; his mane of brown hair swept his shoulders. "I canna believe I am telling you this. For I have never told another soul."

"Your story is your own. It is not for us to repeat," Talfryn assured him.

"I thank you for that." He glanced down at the broken stick. "I thought I was done for," Connell continued. "For you see, I was on the Red Bard's land. He is nay friend to the Roses, and my brother had married one." He angled his head towards his niece. "Fenella's mother. On the road, the full moon shined on me like a beacon. And I could hear the Red Bard's men thundering aboot through the trees."

"What does this have to do with Trooping Fairies?" Talfryn asked.

"Patience, lad." Connell let the broken stick tumble from his hands as he rose to his full height. "A *fairy* appeared at my side—thin, tilted eyes like yours, but his skin was blue violet like the sky at sunset. In a blink, the fairy cast a spell that sent us up into a tree. Afore I could utter so much as a gasp, swords drawn, Red Bard and his men rode beneath us. Once they were gone, the Trooping Fairy, for that is what he later explained he was, made me an offer. If I promised to do three tasks for him, he would see me safely home."

"Three. Sounds like a fairy," Vendel remarked.

"What were the tasks?" Fenella asked.

"First, that I study Norse history."

Talfryn and Vendel exchanged glances.

"Second, that when a niece with flame-red hair was born, I protect her with my life." His eyes glistened. He met Fenella's gaze and placed his hand over his heart. "Och, I failed you. I should have been there to prevent the Blackwoods from stealing you."

Fenella shook her head.

Connell covered her hand with his. "Ever since, I have expected the fairy to reappear and my reckoning to begin. And well I would deserve it since I have also failed to free you."

Talfryn's voice dropped to a whisper. "What was the fairy's third request?"

"That if ever a lad named Talfryn asked for my help, I would assist him."

"By the saints!" Fenella said.

Talfryn raked his fingers through his hair. "Did the fairy give you his name?"

One corner of Connell's mouth lifted. "Ah, he said you would ask that. But 'tis a trick question. Fairies doona give their true names unless they really trust you. And he had just met me. All he said was that his name was somehow related to my clan."

"Uncle Connell?" Talfryn asked.

Fenella shook her head. "Nay. Uncle's name is Connell Thorburn. Our clan includes my mother's family, the Roses."

"Oh! I understand!" Puffball raced out from behind a beech tree.

Connell jerked back; his hand flew to his sword hilt.

Puffball furrowed her forehead, then mirrored Connell's surprised expression. The Scot scowled at her and folded his arms over his chest. Puffball planted her bony green hands on her slender hips and stared up at Talfryn. "Don't you see? Rose. As in blossoms, petals, thorns. *Briar.*"

Realization dawned like the summer sun. "Thor's hammer!"

"I confess, I see nay connection," Fenella said.

Talfryn's skin warmed. *Clearly, thankfully, she has forgotten my transgression when she first met Beacon.* Many a night he had lost sleep worrying, hating himself, for blurting out his father's secret name. *What if Fenella told another human, and that person used Beacon's secret name, Briar, to control him? The fault would be mine. All mine.*

If I can retrieve the Aeternitas Potion, it might make up for my earlier mistake.

"Trust me, there is a connection." His awareness shifted to Connell. "As my pixie Puffball divined, the fairy you met is my father. And I—we—could sorely use your help."

Puffball bowed low in greeting, and Connell gestured an awkward nod in kind.

Then Talfryn explained his quest. As he relayed what the Sight had shown them in the *drabardi's* fire, Connell held up his hand. "Halt."

Baffled, Talfryn stopped midsentence.

"You heard a man reciting psalms over battle cries and the hail of arrows?"

"Aye," Vendel jumped in. "The thaness had closed her eyes by then, but I was watching the *drabardi's* fire. The man was on a Norse warship in the midst of battle, and he refused to fight."

Connell rubbed his jaw. "By the stone! Talfryn, your mother was with a future Christian saint."

"But my father said she believed in the Norse gods and goddesses, especially Frigg and Hlín."

"She may not have converted. 'Twas the Battle of Menai Strait, off the coast of Anglesey. Viking raiders fought two Welsh earls, Hugh the Proud and Hugh the Stout. The battle occurred in the year one thousand and ninety-eight. One of the Christians refused to fight or take cover during the battle. Instead, he remained on the ship during the fray and chanted psalms."

"Was he killed?" Talfryn asked.

"Nae during the battle. But he was branded a coward and soon thereafter fled by swimming ashore. Think, Vendel. Was Disa on the same ship as the man reciting prayers?"

Vendel stroked his sandy beard. "I didn't see her on the same ship."

"Nor did I," Talfryn said.

"I was hidden in Talfryn's pocket." Puffball sat down on the toe of Fenella's shoe. "So, I have nay idea."

Connell lifted his gaze to the loch. "Let us assume she was on a different boat then and had the bad luck of being in the same waters."

"When I saw her in the Finding Horn, her ship was tossed by huge waves," Fenella said.

"Could have ended up anywhere." Vendel pressed his forehead to his blackthorn staff and uttered a frustrated, "Aaugh."

"Father said to search for her in the far reaches of the Northern Kingdom."

Connell scrubbed his jaw again. "The Norse settled in many of the northern islands."

"Talfryn must find her. Queen Mab is holding Beacon spellbound without magic and restrained with Cold Iron. I beseech you, Uncle, help him find his mother and the potion while I search for Mr. Blackwood."

"How can *I* help?"

Fenella detached the Finding Horn from the belt's special clasp, then released the hidden latch. "You have traveled more than any of us." She tilted back the lid with her thumb. "Ask the horn to show you Disa. Mayhap, you will recognize the sea, the land, or even the harbor where she abides."

Connell threw Talfryn a questioning look.

"Please try," Talfryn said. "If Mab thinks five days have passed or if the moon is full in Fairy, she'll enslave, or worse, murder my father."

"Verra well." Connell took the horn, holding it as if it were a baby he feared he would drop. "What do I do with this object?"

"Look into its liquid," Talfryn explained, his heart racing. "Ask the waters to show you my mother."

"Verra well." Connell furrowed his dark, scraggly brows, forming deep creases between his eyes, then stared into the horn's tranquil pool and commanded, "Show me the Viking mother of Talfryn the Halfling."

Chapter Twenty-Eight
Unexpected Allies

1098
The Western Isles

The gods and goddesses have abandoned me. Disa tucked her chin towards her chest. Still, the cold summer rain needled her cheeks as she slogged through the peat bog. Exhaustion weighed her eyelids and slowed her steps, yet she willed her senses to remain alert, her legs to propel her forward. She could ill afford a wrenched ankle from stepping into a hollow hidden among the bog cotton and deer grass.

Disa paused to scan her surroundings. *Nay woods. Nay villages. If I scream for help, only the field mice and whatever beasts might creep down from the hillocks will hear me.* Shuddering, she imagined her gruesome fate should she sink into a deep depression and become stuck.

Her sodden boots squelched in the mud. *How many days have I been walking—how many years? More than eighteen endless years of searching... Since Talfryn's birth...* Picturing her son's adorable fay baby face and surprising yellow hair brought a bittersweet smile to her lips.

Memory of his hastened delivery and her forced departure brought pain anew as rain pummeled the shield slung across her back. The water slid down her padded leather tunic afore soaking into her wool leggings. She held her cloak gathered at her waist to safeguard the vial that had been sewn inside its seam that fateful day. A deep chill not unlike Death's fingers clutched her bones.

"Where are the allies the druid promised me, Ragna?" she called out, her words swallowed by the storm. "Where is the red-haired young woman from the future? *Years* have passed since our time in the cave, seer. I am *in* the future. Where is she?"

The dwarf mage's gift, a tiny crystal sphere given so long ago, pressed against Disa's skin through her pocket, causing a bead of pain like a rock in her boot. With her hands trembling from the cold and rain, she dared not move the crystal for fear of dropping it. 'Til this deluge and long slog, she had had high hopes that soon she would be in the right place, under the right full moon, where she could use the sphere to magick herself back to Grambolyn.

The very thought helped her take another step, then another...

Even so, rain blurred her vision. Puddles dotted the bog, making it difficult to discern fresh standing water from potentially treacherous hollows. The fisherman who had ferried her to the isle had told her to keep the loch behind her and head for the hillocks. He had also warned her of the coming storm. But in her rush to reach the standing stones, she had paid him nay heed. This close to Midsummer, with the long hours of daylight to illuminate her way, surely a brief storm would not hinder her. So she had thought.

Hah! She shook her head, splattering large droplets onto her shoulders and chest. *I am wrong-headed to a fault. Why am I so unable to change course once I set my mind to a particular path?*

Twisting about, she saw only the gray of the tempest. Storm clouds reached the ground, blocking the sun and concealing the now distant loch.

Turning back, she searched the terrain ahead. *According to the fisherman, the standing stones should be located beyond that next hillock.*

"Folks living near the stones may profess to be Christian," he had said, as though assuming she worshiped his one god, *"but they are apt to follow the Old Ways."*

I can only hope so. I will need the aid of someone who believes in magic and fairies.

As she staggered forward, she heard a horrible *squelch* and felt her right foot sink into the mire well past her ankle. "Nay!"

Her nape prickled. She stilled, sensing her fylgja, the spirit bear, behind her.

Have you come to attend me? In an icy rain, on a bog forsaken by the gods and goddesses, so close to the standing stones that I can almost feel their magic? Does Death dare come for me?

"Begone!" she yelled. Struggling to lift her leg out of the muck, she instead sank deeper. The spirit bear huffed cold breaths against the back of her head. "Begone, I say! I refuse to die today. Do you hear me? Today is not my death day!"

"Gothan nott!"

Do my ears deceive me? Dropping the hem of her cloak, Disa twisted, swinging her circular shield from her back to her hand in one fluid movement. As she peered through the rain towards the next hillock, she shifted her free hand to the hilt of her dwarf-forged sword.

Lantern light bobbed in the distance. Driving rain and the gray ground cloud hid the light bearer.

The lantern is held far too high for the carrier to be a goblin. Could a Will o' the Wisp be so tall? She shuddered. *If so, Frigg save me.* She wished she could safely reach the goddess rune nestled beside the crystal sphere in her pocket.

The light bearer approached as though honed onto her.

I carry nay light. How does he know where I am? Disa struggled to free her foot. "Release me!" she wailed.

Just then, the lantern wielder emerged from the ground cloud. Disa tensed as she assessed the imposing figure. His broad shoulders indicated a muscular frame. The rain-soaked cloak he wore could easily hide an axe and all manner of knives.

She tightened her hand around her sword hilt.

As if reading her mind, the stranger stopped and flipped back his cloak to show he bore nay weapons. When she did not lower her hand, he placed the lantern on a boulder, lifted his cloak above his waist, then slowly turned around. Nay weapons were visible. When he faced her again, Disa nodded. The stranger released his cloak, which fell below his knees, covering him once more.

Knowing too well a man of such size, even if unarmed, could overpower her, Disa lowered her hand but not her guard. Besides, the man would be a fool to venture onto the marshlands without at least a hidden dagger.

"I am Torolf," he called out in a welcoming tone. He lifted the lantern so she could see his face. "My wife sent me to find you. We live on the other side of the mound." He jabbed his thumb towards the hillock.

"I have never met your wife," she argued. "So, she could not possibly have known I was here."

"Idony is gifted with Second Sight. We are druids. Farmers. You are a time traveler, já?"

Tingling in the back of her head scrabbled down her nape. "I…" The pounding of her heart rushed her ears.

Up close, she could see he was not the same druid who had spoken through Ragna all those years ago in the cave. She sensed the man afore her practiced the Old Ways but was at heart a farmer. *Mayhap his wife alone holds the magical power.*

"Idony thought you might be hesitant. She would have come as well, but we have two young children." He reached for something inside his cape.

Disa grasped her sword hilt.

Heeding Disa's reaction, Torolf slowed his movements. He pulled a rune from his pocket and held it in front of him.

Disa's breath caught. Instinctively, she pressed her fingers against Hlín's gift, the matching rune secured in her pocket.

"She said this would have meaning to you."

Gaping at the Goddess rune, Disa believed. "What color is your wife's hair?"

His brow crinkled. "Blond."

Not red. Still, the rune proves the woman has Second Sight, and the man braved the storm to find me.

"Rain is abating," Torolf observed. "May I guide you to shelter and a warm fire?"

"I cannot move," she confessed.

Torolf raised his lantern. His gaze dropped to her feet, and he made a sympathetic clucking sound with his tongue. "You are not too deep. Gradually back up."

Disa shifted her weight to her back leg, then froze when mud engulfed her left foot. "I am sinking!"

Torolf circled her, avoiding the quicksand. "Fear not. I know this bog well. Try again. More slowly."

Teeth chattering as much from terror as the icy cold, Disa again shifted her weight. This time, her front leg rose slightly. She sensed the spirit bear retreat as Torolf reached her.

"Hand me your shield."

Remembering how many times she had wielded the shield as a weapon, Disa reluctantly did as he asked. She didn't breathe again 'til she heard Torolf set the shield down a short way behind them.

Torolf reached under her arms. "Lean back. We'll create a counterweight and lift you out."

The bog did not yield easily. But Torolf never wavered, not even when lightning stabbed the sky. Her spirit bear paced. Disa heard the splashes as it padded behind Torolf. She had nay strength left to demand the bear leave.

Ever so slowly, Torolf pulled her free.

"I thank you." Disa stood, too exhausted to move, her quivering legs cold and heavy with mud. Relief loosened the tightness in her shoulders. Frigg had sent her allies! *Thank you, Goddess!* She had had so few allies in her life.

For a moment, Disa imagined Bree's gift, the yellowish-green stone heart, glowed to life and radiated warmth through her pocket. She could almost hear her sister's little girl voice urging her onward. Then, through rain-jeweled lashes, Disa thought she saw Bree with Muirgen bathed in Torolf's lantern light, untouched by rain or time. *Oh, welcome apparition!* Muirgen glanced back her way, his large fairy eyes shining. He hissed and the spirit bear, Disa's *fylgja*, shrank down to a silver cord that spiraled into her belly.

Today is not my death day.

Chapter Twenty-Nine
Faith Tested

"...or worse, murder my father."

Edward heard a string of muffled words but could only make out one vague phrase. Had he heard the deep masculine voice correctly? *Who speaks of murder?* He was still unable to see the man through the trees.

Och! I am surely daft.

Edward's suffocating dread had mounted when he had ridden onto Blackwood land. He had sensed his men's sideways glances. Nay doubt they thought he had lost his mind. Mayhap he had. None knew what had transpired with the *drabardi.* He had ordered the men to continue south towards Argonshire and to keep a watchful eye for Lady Fenella. On the wordless ride, he had nae given up hope. Nae having eaten, nae having slept, he had scanned the road, trees, and hills for Fenella.

Having neared the stronghold, with the stone donjon atop a tall rise just beyond the trees, his faith had faltered. Within the quarter hour, he would have had to face his father and tell his lordship his prized ward, his pawn, had again vanished. While he had worried his hand over his thigh, thoughts of the unopened pouch hidden in his shoe had caused him to question himself.

Has the time come? Should I open it? He had gone so far as to slide his hand towards his ankle, then having righted himself, he had reaffirmed his chosen path. *Nay. I will keep the faith. Nay matter the consequences.* Imaginings of the punishment his father would extract had made him wince.

Then a flash of scarlet among the trees had caught his eye and had caused his pulse to spike.

Once having signaled the contingent to halt, he had swung off his horse. Then he had signaled, again, for his men to remain mounted and silent, having cast a warning glance at Derry, who had already slid off the gift horse's back. Nae having wanted to appear a madman, he had crept rather than run through the trees.

Now, he was near enough to hear a voice, someone speaking of murder. Edward hoped he had misheard, that 'twould be the thaness he had found, and his quest would be at an end. He moved closer to see who 'twas.

Fresh deadfall littered the woods. Quietly as possible, stepping over tangles of branches and leaves, he rounded a beech. His heart stopped, then relief flooded him.

Fenella.

She stood beside her uncle, who was staring into an ale horn. Edward did nae recognize the muscular blond whose back was to him. Brown clods of earth and green bits of Lady's Mantle smudged the stranger's cloak. *My unknown rival.* Edward's blood heated as jealousy flared like a castle fire. He forced himself to inhale long and deep, willing the briny air to quell this unbidden passion afore he stepped within better hearing.

"But I fear I have bad news for you, lad," Connell Thorburn said, looking up from the horn.

The unexpected snap of a twig beneath Edward's boot made

the stranger whirl to face him. Surprise shined in the young man's gray eyes.

Discovered, Edward made quick assessment of the newcomer's yellow shirt, soiled cloak, and intricate Norse wrist cuff. *Is he Highlander, or Norse?* His gaze darted to the blond's waist belt. *Unarmed.*

Intrigued, Edward switched his attention to Fenella's uncle. "Pray, tell us Connell Thorburn. What bad news have you?"

"Mr. Blackwood!" Relief and more tinged Fenella's voice.

Edward opened his non-sword arm. Fenella rushed to his side and wrapped her arms around his waist. He drew her close, his heart in his throat. Her mantle fell to her shoulders, revealing more of her bright tousled hair, and he inhaled its scents of spices, sweet briar, and the sea.

The stranger watched them. Surprise, worry, then loss played across his face afore he turned away.

"You are shivering," Edward said, splaying his hand across Fenella's back. She nodded against his chest.

Connell Thorburn lowered his drinking horn and latched the lid. "Thank the stars you are safe, Mr. Blackwood. My niece has been ill with worry ever since I found her alone on the road."

Fenella pulled away, and Edward's chest tightened. "Are you hurt?" he asked, smoothing his hands down her arms. "We searched and searched. I feared the Red Bard had captured you."

"I am unharmed. Though how we missed each other on the road is a puzzle." She wrapped the edge of her mantle around her scarred hand. "Did Derry and your men escape the monster in the loch?"

"My lady!" Derry darted out of the trees. She halted when she saw the others and lowered her gaze.

Edward frowned at his men, who followed close at Derry's heels. *Dinna I tell you to wait?* They avoided meeting his glare 'til they spied Fenella, then their grim expressions lightened.

The blond paled and backed up as the contingent gathered behind Edward. Thorburn clasped the young stranger's shoulder, as if to keep him from fleeing.

"Who is your friend?" Edward asked Fenella as he brushed a suspicious spot of dirt and Lady's Mantle from her cloak.

"My foster son," Thorburn answered for his niece. "I was aboot to tell him his mother has sent word, calling him back to the Isle of Lewis."

"Lewis? That is MacLeod territory. *Murder is nae out of keeping with that lot.* Why would you foster one of their ilk?"

"Well-formed alliances always have worth."

"Aye," Edward conceded. *But the Isle of Lewis? The ride from Thorburn land to the shore would take at least a week. Then there would be a rough sea to cross.*

Thorburn squeezed the MacLeod's shoulder, turning him so they were face to face. "Safe journey, lad. Wish I could see the Callanish Stones again, and your mother. But Lady Fenella and Mr. Blackwood are to be married on Saint John's Day. And since her father is nay longer with us—"

"I understand," the fosterling said. "I'll see to my mother and then return as quickly as I can." His gaze flicked on Fenella, then dropped. "Ready, Vendel?"

"Aye." Vendel stepped forward.

Edward tore his gaze from the MacLeod and glanced down. "I remember you! You were waiting alongside the road with flowers for the thaness."

"Indeed, my lord. I am here to accompany the lad on his journey. Many blessings on yer upcoming wedding."

"I thank you," Edward's gaze slid beyond the dwarf. He scrunched his brows as he noticed the scattering of beached fish, and the boats rocking in the water. Puzzled, he dragged his attention back to his betrothed and the stranger.

"Thaness." The fosterling bowed his head.

Fenella blinked several times. Dark circles shadowed her eyes. With a twinge, Edward wondered when she had last slept. "Safe journey." Her words sounded heartfelt. "Godspeed."

"Farewell," the MacLeod said in a husky whisper.

"Come." Edward slid his arm over Fenella's shoulder. "We must warm you afore the fire and bring you a hot meal." *Then, mayhap, you will tell me aboot this rival you have brought to my door.*

Chapter Thirty
Difficult Decisions

"Did you see how she flung herself into Blackwood's arms?" Talfryn crossed the beech grove with long, troubled strides.

"We cannot leave without Fenella," Vendel said. By now they were out of earshot and hidden by the beech copse.

"We can, and we will," he avowed. "Did you see how relieved she was to see him?"

"Mayhap it was an act to divert attention from ye," Vendel huffed, hurrying to keep pace.

"It was nay act." *Her heart belongs to Blackwood.* He glanced back in Fenella's direction. *I should be happy for her.* Instead, he felt...unsettled.

Vendel thrust the blackthorn staff across Talfryn's waist, forcing him to stop. "Fenella has studied the Norse. She can blend in among them, and she's immune to Cold Iron. Ye haven't, and ye aren't."

"Glimmer said the serpent armlet would work two more times. I'll be safe."

"For a short time. Remember Jorvik? What if the magic ends afore we find Disa?"

"The enchantments will last long enough. Besides, Stanka said her ring offers protection."

"Should we trust her?"

"Should we trust any human?" Talfryn squatted and ripped a handful of moor grass from the ground, then threw it. "When Blackwood spoke to you, Connell leaned close to me and whispered a new date. But I didn't have a chance to ask him about the Callanish Stones."

Vendel shook his head. "Never heard of them. Mayhap we should waylay Connell and bring him."

"He'd add too much weight. We would have crashed into the water on our way here if not for—"

"Lily. Lily saved us."

"Aye." *Lily.* At the thought of what she had risked, his heart expanded; his Highland shirt suddenly felt too tight across his chest. Talfryn cleared his throat. "Lily must have taken the equistragon without permission since my father was incapacitated and Glimmer was away."

"Members of the Noble House of Storms and descendants of the Hawthorn Throne have tamed, ridden, and mastered the beasts long afore your godmother appointed herself Supreme Weather Councilor." Vendel scoffed.

"But Glimmer is in charge now, and Lily chanced her future to save us." *To save me.* Unchartered emotions warmed his core.

"Well, we'll have to make do on our own now." Vendel secured the blackthorn staff in his waist belt. "Finding your mother will not be easy."

"Why did my mother tell Father not to search for her? Was she trying to protect him?"

Vendel gestured to Talfryn to bend down so he could whisper in his ear. Talfryn strained to hear him over the waves and wind. "She feared he would lead Mab to the rightful heir of the Blessed Court throne."

"But there is nay—"

"'Tis not safe to speak further. I may have already said too much." Vendel angled his head, directing Talfryn to a fairy hollow at the base of one of the beeches. "As ye suggested, we will journey light. I pray to the gods and goddesses for a positive outcome." The dwarf extended his arms.

Dazed, Talfryn clasped Vendel's hands. Glancing at the fairy hollow afore returning his gaze to Vendel, he thought his intended destination instead of speaking it out loud. *To my mother, Disa. Callanish Stones. Isle of Lewis. The year one thousand and ninety-eight.*

White light exploded around them, exposing a surprised woodland fairy's face peering out from the hollow. Talfryn's stomach whooshed to his throat as a powerful force whipped him off his feet. He teetered, thrown off-balance by the weight of his Norse muscles and Fenella's absence.

Talfryn closed his eyes and let the wind rush his ears. The centuries fell away, and the sounds of human endeavors grew faint.

His mouth watered when they flew above cook fires and he smelled roasting meat. Then the air chilled. Talfryn's feet grew numb within his thin leather shoes, and his skin dampened as he and Vendel rushed over misty lochs. Thrice he heard fairy music. A longing for home and his father's freedom entwined his heart. The faint rhyming chant of a wizard reached his ears. *How much farther?* he wondered. His mouth grew as parched as month-old herbs hanging in the drying shed.

Talfryn's stomach rushed up to his throat as the magic dipped Vendel and him beneath one of the ominous gray clouds that had overtaken the sky. Soon, the roar and crash of enormous waves drowned out the rushing wind. The temperature plummeted. Seawater writhed below, and bobbing in the turbulent swells, fierce-faced storm kelpies shook their fists at them.

"What is their grievance?" Vendel asked.

"Nay idea."

A few kelpies waved their blue-gray arms, as if motioning them to turn back.

"We must proceed!" Talfryn shouted.

Three kelpies assumed their horse form and slapped the waves with their tails, creating thunderous booms.

"Hardly welcoming." Vendel frowned.

"I thought kelpies only haunted rivers," Talfryn shouted over the thunder.

"My belief, as well. Mayhap 'tis different this far north." Staring at the churning sea, Vendel's ruddy face grew ghostly, then green.

"There's an island ahead." Talfryn tightened his grip on Vendel's clammy hands. The vastness of the isle gave him pause. Forests, contained largely but not exclusively to the south, yielded to vast tracts of rolling moorland. Much of the island's interior appeared uninhabited. *Good. Less area for us to search.*

The magical force continued its northern course, then banked westward and pushed them across the middle of the island. As the kelpies' thunder faded, Talfryn scanned the rocky hills and undulating moors below. Wind ruffled his Norse clothing. He spotted scatterings of alder, birch, hazel, and rowan.

Without warning, the magic gathered speed.

"The west coast is ablaze!" Vendel exclaimed as they whirled towards the conflagration.

Talfryn peered at the flames dotting settlements along the western seaboard. "Balefires?"

As the words left his mouth, screams and smoke, carried on the wind, hit them full force. Talfryn glanced at Vendel and saw his own apprehension reflected in the dwarf's wrinkled forehead and pinched eyebrows.

"Norse invaders." Vendel's words sent dread coursing through Talfryn's veins. The dwarf dipped his chin as they flew over the southern tip of an inlet. "Look! Dragon warships."

"I thought the Norse had settled here." Talfryn swallowed the fear constricting his throat. "Why are they attacking their own people?"

"Nay idea. Power? Land? Revenge?"

Talfryn uttered a Change Course spell, then repeated it when the first incantation failed. "I can't make the magic veer away."

White vapors coated his eyes and shivered his skin when they sailed through a low, wispy cloud. Flexing his toes as the soles of his shoes skimmed treetops, he blinked to clear his vision. "Mother must be down there. Somewhere." *She's a shield–maiden. Is she attacking or defending the island?* As they neared a cluster of burning dwellings, Talfryn lifted his legs to avoid the heat rising from blazing roofs and squeezed his eyes shut against the sting of black smoke. Thoughts of Lily and her singed wings seared his mind and gave him pause; he hoped he would avoid a similar fate.

Hearing the angry crackle of voracious flames, Talfryn forced his eyes open. *Mab's tears.* Far below him, humans and livestock ran for the hills like deer fleeing a forest fire. Tree spirits leapt for safety, overwhelming a nearby hazel when a copse of sacred rowan and birch combusted.

Someone cursed the name Magnus Barelegs.

"Where are the Callanish Stones?" Vendel shouted.

Talfryn squinted through tears welled by the smoke. In the distance, he spotted a dense forest. Scant human dwellings dotted the area. Nay fires. The torch-bearing plunderers must not have advanced that far north.

"There! Beyond the woods! Standing Stones!" Talfryn angled his head. Not just a stone circle, as he had expected, but a vast complex laid out like a skewed Celtic cross or an enormous dragonfly. From his aerial vantage point, he saw other stone circles close by. *Which set of standing stones is the right one?*

"Where?" Vendel asked.

"Over there! Has the wind blinded you?"

Ahead, a ring of magic shuddered off the standing stones. Primeval smells, wild and mysterious, rode the nearly invisible spell wave. Ancient drumbeats pounded against Talfryn's chest. And frenzied chants embedded in the magic swelled as the first ripple neared. Talfryn's flesh crawled.

"Brace yourself!" Talfryn warned. The spell surge crashed against him, pummeling the air from his lungs. Its force shoved him backwards. A moan pried open his gritted teeth and escaped as his legs flew up, and he narrowly missed kicking Vendel.

Then, as though the stones summoned the magic back, the spell wave receded, drawing Talfryn and Vendel within its wake. Talfryn's stomach whooshed.

Vendel shook his hair out of his eyes. "That wasn't too awful."

Talfryn eyed the megaliths. His pulse quickened. He tensed and tightened his grasp on Vendel's hands.

"Not another—" The second spell wave cut off Vendel's words.

Magic slammed Talfryn's chest like a giant wielding a tree trunk as Vendel slipped from his grip with a yelp. When the dwarf plummeted, the blackthorn staff flew out of his waist belt and sailed off.

Talfryn's ears buzzed. Every organ hurt; every muscle screamed. He fell like a boulder dropped off a cliff, falling fast and hard. It seemed the ground was rushing up at a dizzying speed. Words to spells blurred in his mind. All he could think of was Luminous.

"Summon her!" Vendel cried as he tumbled through a smoke drift.

"Luminous!"

The air thrummed with the *whap, whap, whap* of colossal wings. Luminous galloped out of a sun-rimmed cloud. Sunrays glanced off the great steed's silver hooves while wind flared her mane and tail. She swooped beneath Talfryn.

Talfryn thudded onto Luminous's back. "Save Vendel!" he commanded breathlessly, reaching forward to snatch hold of Luminous's silver mane.

The shimmering winged horse lowered her head and dove through the sky, swooping below, then upwards to catch the dwarf as he fell. Vendel landed behind Talfryn and clutched at the Halfling to gain a firm grip. The white steed continued to soar upwards on a wind stream, gradually slowed to a level glide, then pranced across a low cloud afore descending and gracefully alighting on a peat bog. She snorted and bobbed her head while Talfryn and Vendel slipped shakily from her back.

Talfryn threw his arms around Luminous's neck. "I don't know how you found us! Thank you!"

Turning her head, the winged horse fixed a pale, knowing eye on him. Talfryn reddened under the mount's withering appraisal and dropped his arms to his sides.

"You have been following us!"

Luminous bobbed her head.

"Such good fortune!" Vendel placed his hand over his heart. "Vendel Fairheart. 'Tis a pleasure to meet ye. I thank thee for rescuing me."

In response, Luminous lowered her head and sniffed the dwarf, then nickered.

Talfryn twisted his father's ring and scanned their surroundings. He saw nay fire, detected nay hint of wind or smoke. The crevice-ridden moor spread out in all directions. "I've lost my bearings. Where are the standing stones?"

Vendel glanced about. "I am not certain. I never saw them."

"What do you think, Puffball?"

When no shrill voice answered, Talfryn threw back his cloak and clutched at his pocket. *Empty!* Fear rushed him. He threw Vendel a wild look. "Where is Puffball?"

"I haven't seen her. Did she fall out?"

Remembering the storm kelpies in the writhing sea, Talfryn grew queasy. He cupped his hands around his mouth and yelled, "Puffball!"

"Did we leave her behind?" Vendel asked.

Talfryn dug his fingers into his temples.

Vendel pointed the staff at him. "Ye do not know because of Blackwood!"

The truth stung. Anger flushed his skin—anger at his parents for stealing the potion, anger at Blackwood for stealing Fenella's heart, but mostly anger at himself for stealing away without his precious pixie. *I'll never forgive myself if Puffball has perished.*

"Puffball *must* be with Fenella." Talfryn rubbed the back of his neck. "I am almost certain of it. One of us would have seen or heard her if she fell out. Don't you agree?" Without waiting for Vendel's answer, he rambled on, trying to convince himself as much as the dwarf. "Either way, she's a pixie. She's not defenseless."

"Can she swim?" Vendel asked.

"Nay."

"Fly?"

Queasiness in Talfryn's stomach worsened. "Fine. She cannot fly, only leap. But Puffball is highly resourceful. We will have to trust she is well 'til we can find her. She would want us to retrieve the Aeternitas Potion and save Beacon."

"She would want us to keep our eyes and ears open for her."

"Every second," Talfryn vowed, painfully aware of the rectitude in the dwarf's rebuke. "Let's find my mother, then be quit of this place."

"Try the Finding Horn," Vendel suggested.

Talfryn reached for his waist belt. His shoulders drooped. "Connell still has it."

"Have anything else? A diviner's stone? Food?"

"You toil in the crystal mines. Mayhap *you* should have brought a divination ball."

"Do ye know how rare diviner's stones are?" Vendel brandished the blackthorn.

Talfryn's cheeks heated. "Aye. They are quite rare. My apologies. I am the one who failed, not you." *I've failed everyone. Nettle held captive, Puffball missing, Father still endangered, Lily and Fenella at risk… If any of them die, the fault will fall heavily on my shoulders.*

Vendel retrieved the Luminositor from his hidden pocket. With a twist, the small device whirred open and revealed a bright miniature moon.

Talfryn's heart sank.

Vendel retracted the gadget. "We best hasten."

Chapter Thirty-One
Lost and Found

The Blackwood Stronghold, Scottish Lowlands

"We must make haste. Lord Argonshire awaits." Mr. Blackwood held his hand firmly against the small of Fenella's back as they approached the horses. "Ride with me. I will nae risk losing you again." Blackwood's breath, warm against her ear, fluttered Fenella's core and brought a flush to her cheeks. He mounted first, then extended his hand. Blackwood's strength pulsed through her as he swung her up and onto the saddle behind him. She became keenly aware of her body pressed against his and of the heat issuing through his mantle. Uncertain where to place her hands, she rested them on her thighs.

When the contingent approached the narrow passage leading to the portcullis, Blackwood reached back to draw her arm around his waist. "Hold on. The way is steep."

Fenella slid her other arm around him and flattened her cheek against his back, his heartbeat steady against her ear. *I am sorry I lied,* she thought. *Please forgive me.*

The clatter of the horses' Cold Iron hooves, along with the groan and squeak of the gate being raised, filled the passage.

Fenella glanced up at the high walls, and her stomach clenched as she felt the full ramifications of her return to a captivity that had barred her from her mother's deathbed, kept her from her duty as a daughter, and robbed her of a final farewell. As they passed beneath the iron gate, Fenella flashed on her mother's chamber, devoid of life and joy. *Your shrouded mirror will never again contain your bold reflection, nor capture the warmth of your affectionate countenance. You died without me there to hold your hand.* Anger, like a festering wound, sprouted anew.

The horse halted, curtailing her descent into fury. She felt Blackwood shift position, and the saddle creaked as he stretched his leg in preparation to dismount.

"Niece?" Uncle Connell had already dismounted and was extending his hand.

"Aye." Fenella scooted backwards and brought her leg over the horse's back. Connell clasped her waist and lifted her down. She watched him glance at Blackwood, who had his back to them while he dismounted and spoke to a groom. When Connell's eyes met hers, he pressed the cool, hard surface of the Finding Horn against her palm. Instinctively, she curled her fingers around the magical object, then secured it to her waist belt.

Blackwood turned their way, then glanced up at the stronghold. The gold flecks in his hazel eyes darkened. "Shall we make our presence known to his lordship?"

Fenella crossed her arms over her chest. "I have nay words to say to him."

Her uncle drew her close and whispered in her ear, "Keep your enemies near." He plucked a beech sprig from her tangled hair and dropped it to the ground. "Let the lass right herself first," he told Edward. "She has had a difficult go of it."

Derry sidled up to her, wordlessly offering assistance. Angus was nae far behind. He appeared uncertain and stood tight-lipped, awaiting orders.

Edward hesitated.

"Look, 'tis Calidore!" Derry waved at the mountainous Scot, then smoothed her hair as he descended the outer timber stairs.

Fenella watched and waited as Edward's glance traveled from Derry to Calidore, then back. A sudden smile brightened his face.

"Angus." Stepping forward, Edward clasped the man's forearm. "I thank you for your service." He raised his voice to include all his men. "I thank you all. You have earned a good meal and time with your families. See to your horses, and I will meet you in the Great Hall."

Trustworthy Calidore joined them, and Fenella caught the relieved expression on the guard's face afore he masked it.

"Calidore, escort the thaness to her chamber," Edward said.

"Aye, sir."

Derry quietly clapped.

Angus averted his gaze, then strode back to his horse.

"I thank you, Mr. Blackwood." Fenella nodded to him and her uncle, then took her leave as Derry fell into step behind her. Calidore loomed above both like a protective shield, keeping close as Fenella made her way towards her chamber in the east tower.

"I sent a lad to light the fire," Calidore said.

"That is most appreciated, Calidore."

The scent of roasting beef wafted from the kitchen, suggesting a hearty supper was being prepared. She doubted any of Edward's men would tell his lordship about her disappearance. But would Angus hold his tongue? Mayhap, since the incident reflected poorly on his skills… Still, would Argonshire confine her in the tower without food if he heard of her transgression? Or would he devise a worse punishment?

Unease washed over her.

"Derry, take time to see to your own needs. Then fetch me hot water so I can wash the journey from my skin."

"Aye, my lady. Right away." Derry glanced up at Calidore, then veered off to her own quarters.

Little light angled through the high, narrow windows in the passageway. Fenella hastened through Argonshire's chamber and Gavin's quarters, then moved quickly up the stairs, Calidore ever at her heels. She reached her chamber door ahead of him. The portal was unlocked, so she slipped inside and closed it behind her, found the key, then locked it.

Her skin prickled from the cold, even though fresh rushes had been strewn on the floor and the promised fire crackled and snapped. Her curtained four-poster bed, with its thick furs and quilts, beckoned. Without removing her cloak, Fenella unfastened the Finding Horn from her belt and hung it on a hook carved into the bedpost. She pulled the privacy panels closed, then sank down onto the bed.

"You can come out now."

With a scratching noise, Puffball emerged from Fenella's boot.

"You should throw away your dagger," the pixie huffed, rubbing her spindly green arm. "The jewels on the hilt poked me."

"Sorry. But you are lucky I dinna scream when you climbed in. I thought you were a rat."

Puffball collapsed on the top fur and closed her eyes. "Is it my fault Talfryn strode off afore I could hop on his boot?"

"Of course, nae." Fenella had a new worry. "Will you be unwell inside the castle with all the Cold Iron? I dinna consider my dagger's blade while you rode within my boot."

The pixie sat up and sniffed the air. "There's no Cold Iron here."

A blush flamed across Fenella's cheeks. "I tossed all but my dagger."

Puffball grinned. "I knew you liked Talfryn!"

Her blush deepened, scorching her pale, freckled skin. "But the rest of the stronghold is bursting with Cold Iron."

"Pixies are not afflicted by the metal. Even so, I'll hide here 'til we come up with a plan."

Fenella nodded in agreement, then stared at the Finding Horn. "I wish Uncle Connell had had a chance to pass the horn to Talfryn. He'll need it to find his mother."

"Well, the Lady of the Lake's handmaiden gave it to you. So, I think you're supposed to possess it. Look and see if you can find Talfryn and Vendel."

"The Blackwoods are waiting. And Derry will be here any minute."

Puffball snorted. "Let them wait."

Chapter Thirty-Two
Choices

The Isle of Lewis

"Wait a moment, Vendel." Talfryn licked his finger, raised his arm, then swiveled 'til he felt a faint breeze brush his fingertip. "The smoke we saw from the air billowed west to east. So, the sea must be in that direction." He pointed. "Mayhap we are too far away to smell the smoke."

"Then, the standing stones are north?" Vendel asked.

"Aye." Though clad once more in the itchy wool tunic and sea-blue breeches of the Norse, he still wore his fairy cloak and began to rummage through its pockets. "But how do we find the right stones?" He discovered the sacks of pixie and fairy dust Puffball had given him. Fresh worry tightened his chest, and he swallowed hard. *Puffball is still alive. She has to be.* Then his fingers folded around his father's gift. Hope rose as he withdrew the silken pouch and cupped it in his palm.

"Beacon's magic dust," Vendel observed.

"He said use its contents…but didn't say how." Talfryn untied the blue-violet pull strings, warily reached in, and pinched a few glittering specks between his finger and thumb. The crystalline granules sparked, then ignited. When heat singed his skin, Talfryn recoiled and tossed the now scarlet flaming specks into the sodden bog. As soon as the magic dust hit the water, it appeared to extinguish.

"Ye wasted it!" Vendel admonished.

For a heartbeat, Talfryn, too, believed the magic was lost. Then his father's scent—sea breezes and spring gardens—filled the air. Glinting embers rose from the bog, flew above his head, circled like a whirlwind, then showered the ground in front of Luminous. A glowing orange path stretching towards the horizon materialized.

Luminous puffed air out her nostrils.

"Tie the pouch and put it in yer pocket." Vendel demanded. "Make haste!"

Talfryn fumbled with the cords as he moved with a sense of urgency. "Step onto the path afore it disappears."

Magic sparkled like hundreds of tiny suns, illuminating the path against the patchy green moor. When he stepped onto it, Talfryn's feet tingled through his soft leather boots. The enchantment warmed his flesh as it crackled around his ankles. Vendel tapped the ground with his staff, and the magic, like a bright whirl of orange autumn leaves, spiraled up the blackthorn. Talfryn glanced over his shoulder and saw the enchanted dust twinkle and dance with each step Luminous took.

Talfryn noted the absence of glens, birds, and forests. "Nay fairies."

"Not anymore." Vendel frowned. "Just peat-covered rock dotted with heather."

"Keep your wits about you. There is powerful sorcery afoot."

"Aye…and torch-wielding Vikings." Vendel grimaced.

"Hopefully, they won't come this far inland." Then an unnerving notion occurred to him. "Do you think my mother is one of them? Is she setting farms ablaze?" The thought made him nauseous.

Vendel halted. "How can you think such an abomination? Disa would never burn families out of their homes. That's how slave traders captured her and her sister. 'Twas Vikings killed her parents!"

The sick sensation in Talfryn's stomach doubled. "Father never told me."

"Beacon was waiting for the right time." Vendel seemed vexed. "Then time ran out."

A slave. He had been too captivated at the sight of his mother in the *drabardi's* fire to absorb all she had said.

With renewed resolve, he stiffened his spine. He studied their surroundings. Having traveled through time, he was unsure if it was still June, if they were close to the summer solstice. If the season was unchanged, then hours of daylight could lie ahead. They dare not fly or otherwise magic themselves to the standing stones without drawing attention. "I wish there were trees for cover."

"Agreed." Vendel retrieved the Luminositor from its hiding place, then twisted the silver axe blade. The crystal disc slid into view. "What do you think?"

Talfryn bent to study the enchanted moon crystal. "A day may have passed. But that would be wrong. Time moves more slowly in Fairy."

"Mayhap magicking back and forth through the centuries harmed the device." Frowning, Vendel retracted the disc, then hid the contraption. "We've halted long enough."

They set off. Talfryn shortened his stride to match the dwarf's. He fixed his gaze on the ground flanking the path. His pulse jumped when he spotted a splash of bright green. But just as quickly, his hopes shattered.

"A mere leaf." Vendel frowned.

"We'll find her," Talfryn vowed. He sniffed the breeze for Cold Iron or the telltale scent of ancient spells.

"Smoke," Vendel said.

"Aye." Talfryn's throat burned. Although the smoke carried nay stench of dark magic, his thoughts nevertheless drifted back to the fire that had harmed Lily and killed her sister. *When a fairy kills a fairy, he or she loses immortality. But I must find and return the one potion that restores immortality to save my father.* His empty stomach churned. *If I do, Merrybelle's murderer may never be discovered.*

As though Vendel had read Talfryn's thoughts, he said, "We have difficult choices ahead."

Chapter Thirty-Three
Secrets and Unknowns

The Blackwood Stronghold

A knock on the door catapulted Fenella's heart into her throat.

"Thaness?" Two more quick knocks followed Derry's quivering voice.

"Pox!" Fenella snapped the lid shut on the Finding Horn and latched it. "One moment," she called out.

"What did you see?" Puffball asked in a low voice.

"A glittering orange path," Fenella whispered. "Conceal yourself!"

Puffball scrambled into the pocket of Fenella's fairy cloak. "Hang it on the peg," the pixie said. "Hide in plain sight!"

Quickly, Fenella returned the Finding Horn to the hook on the bedpost and covered it with a privacy panel, then hung the fairy cloak beside her scarlet mantle on the peg by the door. Expelling a nervous breath, she opened the door.

Derry bent and lifted the bucket of steaming water she had placed on the floor. "Were you talking to yourself, my lady?" she asked as she carried the water over the threshold. In her haste,

Derry brushed against the fairy cloak. The garment fell with a small thud among the rushes.

Fenella caught Calidore's amused gaze as she closed the door.

"Och, sorry!" Derry placed the wooden pail on a small bench near the fire, then recovered the cloak and brushed cowslip and lavender from the shimmering cloth.

"Leave it be." Fenella grabbed the garment from her maid's hands, then, as she replaced it on the peg, in a lighter tone, said, "Nay harm done."

Puffball pulled the pocket open and winked. Assured she was unharmed, Fenella shooed the pixie back inside, patting the cloak as though she were smoothing the cloth.

"Thank you, Thaness." Derry retrieved the bucket, then poured the heated rosemary water into a bowl. "'Tis such a relief to see you safe. I feared the loch monster had eaten you."

"And what did Mr. Blackwood think?"

Derry glanced at the door and lowered her voice. "Och, he was terrified the Red Bard had stolen you. He said someone had kicked him in the stomach and then dragged you away. Is that what happened, my lady?"

"I am nae ready to speak of it," Fenella stalled. "Lord Argonshire and Mr. Blackwood await me." She stripped down, her anger rising anew as she remembered how Talfryn had hurled magic at Edward.

Derry waved her hand over the bowl as if to cool the water, then offered Fenella a chunk of soap.

The warm scented water calmed Fenella's nerves. She quickly washed her face, neck, and arms. *Think! You must offer an explanation to Lord Argonshire. But what rational behavior would account for such irrational events? For magic?* Her thoughts collided.

Derry handed her a drying cloth. "We searched the woods, and then we encountered your Romani."

Fenella stilled, the cloth at her throat. "You did?" Goose flesh rose on her arms.

"Aye. I did nae fully trust them; though they seemed to know you."

"I have allowed them to cross Thorburn land," Fenella explained. *What did Stanka tell Edward?*

"They assured Mr. Blackwood the Red Bard had nae taken you."

"Did they?"

"I am sure of it. Otherwise, Mr. Blackwood never would have ordered the men to continue south." Derry pulled a clean léine from a wooden chest at the foot of the bed. "Here, my lady." She handed her the gown.

Fenella tugged the daffodil-yellow léine over her head and brushed her fingertips across the bright gold embroidery at the neck. She glanced at Derry, and for a fleeting moment saw the maid stare back at her with tilted eyes.

"Something amiss, Thaness?"

When Fenella blinked, Derry's eyes looked normal again.

"Could you find me a comb? I seem to have misplaced mine."

"Of course, my lady." Derry bobbed her head, then darted out of the chamber. Fenella closed the door.

"Puffball?" she whispered.

The pixie poked her head out of the cloak pocket. "What do you suppose Stanka told him?"

Fenella thought back to her reunion with Edward. He had appeared cold and suspicious of Talfryn, but he had treated Connell the same way upon their first meeting. Her stomach fluttered when she remembered Edward's lengthy embrace.

"I doona think Stanka betrayed us. But with Argonshire controlling my estate, she might have pressed her advantage in some other way."

With a whoosh of air, Derry flung open the door and rushed in wielding a carved comb. "Found one," she said.

Puffball disappeared inside the cloak pocket.

Fenella sank onto a stool.

Derry hunched over her and clawed the comb through Fenella's tangled red mane. "Mr. Blackwood was so relieved to see you."

"And I him." Fenella knotted her hands together.

The maid raked through a tangle. "He was nae himself after meeting with the Roma." Out of the corner of her eye, Fenella saw Derry's hand pause midair, clutching the comb. "Imagine! You are to be married in a sennight." Derry's voice sounded wistful. "Such a happy ending to a sad beginning."

Happy ending? Fenella's breath caught in her throat. *I have betrayed Edward by aiding Talfryn. Should he discover the truth…* She lowered her gaze. *Happiness can be stolen in an instant.*

"Oh, my lady, I nearly forgot!" Derry retrieved a familiar object from her pocket and reverently handed it to her. "I found this on your horse and safeguarded it."

"My mother's prayer book!" Fenella hugged the tome to her heart.

Derry's sunny expression collapsed into one of uncertainty. "Mayhap I should have given it to Mr. Blackwood?"

"Nay, I am pleased you safeguarded it. Your faith in my return shall be rewarded." She angled her head towards the chamber door. "My tangles can wait. Take your leave and enjoy the day."

"Thank you, my lady!"

Fenella barely noticed the maid gather up the bucket and slip silently from the chamber. Eyeing the dear memento, Fenella's gaze was drawn to something wedged between the pages. *How distraught I was to have missed this afore.* Her fingers trembled as she extracted a folded letter. Her breath lodged in her throat when she saw her name written in her mother's careful hand. Eyes tearing, she turned over the letter and broke her mother's wax seal.

Dearest daughter,

Much has been asked of you as the Thaness of Thorburn. I am proud of how well and bravely you have born the burdens of leadership. After careful thought and prayer, I feel compelled to add another task for you to shoulder. Forgive your gran. She has nay memory of the day our lives so horribly changed—nay memory of the harm she caused us. Her mind has become locked in the distant past. Her days are spent searching the stronghold for her long-lost parents, and her sister, Esme, who she yet believes was stolen by fairies.

The ominous presence of your captor's men further upsets her fragile state.

I beseech you to forgive her, as I have. Do this for me, my love, and more importantly, for you.

With love,

Your mother.

Fenella read the missive twice, then returned it to the safety of the prayer book, where her cascade of tears would nae blemish her mother's final wish.

Chapter Thirty-Four
Repel Spells and Ancient Whispers

The Isle of Lewis

"The wind has changed," Talfryn said, between coughs. "It is coming overland from the north." Billowing black smoke wafted towards them from the direction of the Callanish Stones.

"How did the invaders move ahead of us?" Vendel wondered aloud as they trudged forward.

Upon reaching higher ground, both looked west towards the sea. Low distant hills sloped downward, revealing a glimpse of the coast.

Icy tentacles of dread encircled Talfryn's throat. *More warships.*

Vendel uttered an oath under his breath.

Talfryn studied the terrain to the north. "There must be farms on the far side of that hill."

"The dwellings we saw from the air—"

"If my mother is truly near the Callanish Stones, she is in danger." Talfryn quickened his pace to a trot.

Luminous whinnied and bobbed her head.

"Yer winged friend wants ye to test the air for Repel spells," Vendel said.

Talfryn stopped and conjured a weighted sack. "I developed this during my banishment to the Borders," he explained afore uttering the enchantment. When he declared the final word, timid tendrils of vapor curled upwards from the sack, forming themselves into dark mice-shaped puffs of steam.

With a hiss, the vapors condensed, then exploded and scattered into clouds resembling ethereal storm-gray cats. The creatures prowled towards one another from the four directions, each emitting guttural meows. As they approached striking distance, their yowls morphed into growls, then grumbling thunder boomed forth as they coalesced around Talfryn's shoulders.

Startled, Vendel jumped, then ducked, while Luminous shied away several steps.

With claws outstretched, the cloud cats pounced, converging into a raging tangle. The sack trembled in Talfryn's open hand as the sky creatures fought in a frenzied circle, creating a vertical wind tunnel that scattered the approaching smoke.

"What are ye about?" Vendel shouted over the whirling winds.

"Testing the Repel spell!" Talfryn braced his left wrist with his right hand as the vortex whistled downward. "Hold onto your blackwood staff!"

With an audible gulp, Vendel clasped the rod in both hands and held the shuddering staff close to his body.

Luminous drew in her wings and leaned into the wind.

With great precision, the narrow vortex whistled downwards and snatched up the pouch. Stung, Talfryn lowered his hands and watched the small sack whirl upwards within the cone as the brawling cats drew the cycling winds skywards. The pouch reached the top of the vortex, and he heard a loud thud as it hit an invisible barrier. Hissing and yowling, the cloud cats broke from battle as they and the wind funnel vanished. The weighted sack plummeted, causing a small thump when Talfryn caught it. With a snap of his fingers, the pouch disappeared.

As the wind anomalies around them settled, smoke, wafting in level with the hilltop, again choked the air.

Talfryn released a long breath. "If we fly low, we may make it," he surmised as he mounted Luminous. "The smoke will provide cover."

"Give me a hand up," Vendel said. "And I thank ye, Luminous, for yer offer."

"Do you read everyone's mind?" Talfryn asked, bending low to clasp the dwarf's hand and then hoist him onto Luminous's shoulder.

"Not women's," Vendel confessed as he squirmed to gain purchase. "Though not for lack of trying."

"Mind the blackthorn," Talfryn said, adjusting the rod so it would not dig into Luminous or stab him. Then he bent his knees and pressed his long legs to the magical beast's sides as he clasped onto the ridges where her wings sprouted from her body.

Vendel leaned forward and clutched Luminous's mane when the horse sprung into a gallop. Her silver hooves sparked against the magic path like stars twinkling in the night sky.

Talfryn tensed as Luminous's muscles bunched beneath him, and with a great heave, she leapt skyward. *Whap!* The horse spread her massive wings, and the air whipped.

"Fly towards the hill!"

Luminous flew with increasing speed, lifting them just above the choking smoke.

For a fleeting moment, the black vapors parted and Talfryn

spied distant standing stones visible through blurry waves of air. *The Callanish Stones!*

He listened to the mumbled whispers emanating, he realized, not from the stones, but from the spell current. Then drumbeats, he sensed had been pounded out by spell-casters long absent from the visible world, thrummed his chest. Shimmering air thick with dragon's blood and meadow rue burned his throat.

The ring of protection appeared even more formidable than afore. Twice his height, the barely visible wave sped through the sky towards them.

"Luminous, dive!" Talfryn ordered.

"Too late. Brace yerself!" Vendel warned.

Head down, neck muscles bulging, Luminous plowed into the otherworldly magic. Her impact bent the spell-force inward, back towards the distant stone circle.

Talfryn held his breath as they stretched the wave's boundary. The drumbeats ceased. Time froze.

Around them, the spell-force heated like a rock baked by the summer sun. Near its core, a murmur rose. Talfryn strained to discern the whispers, but the voices were too distant and quickly morphed into a hum. The drone faltered afore finding its rhythm. Then a chant rose, steady at first, the words maddeningly foreign and incomprehensible. The chant grew faster, louder, like the casting of a powerful spell. Magic pressed against Talfryn's chest. *Are these voices an echo through time of those who set the stones?*

Luminous trembled. Talfryn felt her strain against the enchantments that flattened her wings to her sides. Thunder, loud and terrible, rumbled the sky.

"Notttt goooood." Vendel said.

To Talfryn's ears, the dwarf's words sounded slow and stretched, as if they had traveled through another realm to reach him.

Time restarted. The wave shuddered, then flung them backwards, whipping them through the sky like Thor's deadly hammer, and away from the stones. Tumbling sideways, Luminous thrust open her wings only to have them forced against her sides once more. Talfryn heard Vendel scream as the threesome plummeted through black smoke, not knowing when or how to brace for impact.

As they neared the ground, Talfryn spotted the red flare of a large fire and realized the hill was all too close.

"Lumi—"

With an unearthly screech, Luminous forced her wings open, then abruptly tilted in the wind. The blackthorn staff slipped from Vendel's waist belt and flew off. With tremendous effort, Luminous righted herself. And afore Talfryn could release the breath locked in his lungs, the winged horse landed. Hard.

Momentarily stunned, Talfryn at last pushed up from Luminous's back. His arms trembled.

Vendel relaxed his grip on the steed's mane, then slowly sat up.

"Luminous, are you hurt?" Talfryn gingerly slid from her back. He stumbled but caught himself. Then he reached up to help Vendel dismount.

A rushing, piercing sound made Talfryn and Vendel flinch as the blackthorn staff arrowed to the ground close to Luminous's wind tossed tail. The winged horse neighed and skittered away as the staff struck hard peat. Stiff in stature, the rod teetered, then fell sideways.

Vendel steadied himself afore hastening to retrieve his staff.

Noticing Luminous quaking, Talfryn bent and examined her cloud-gray legs from knee to hoof.

"Isn't she immortal?" Vendel asked over his shoulder.

"No one knows." Talfryn straightened and patted Luminous's

neck. "Nothing appears broken. Well done, girl. You brought us to the foot of the hill."

"Aye, and the safe side," Vendel added.

"We'll catch our breath afore we face whatever is on the other side." Talfryn gathered his fairy cloak tightly around his body. He flashed on a familiar place where the warm evenings coaxed sweet perfume from night-blooming meadow bonnets and long-winded stories from wakeful tree spirits. *Fairy.* Then quick as a will-o'-the-wisp, his pleasant thoughts turned to Mab's painful threat. *By now, Father might be dead.*

"It should have broken, falling that hard and fast. But it didn't." Vendel's assessment as he returned with the rod interrupted Talfryn's ruminations.

"The *drabardi* must have seen a powerful future in your palm to give you such a staff."

"Hmm, well I..."

Whatever else Vendel might have said was lost when a hell-fire roar, punctuated by hungry crackles and snaps, shredded the silence.

Talfryn whirled to see where the danger lay, then ran past Luminous. The glittering path had vanished along with his father's sea breezes and spring gardens scent. Towering flames lit the sky.

CHAPTER THIRTY-FIVE
ALL IS ILL

As Talfryn rounded the hill, a panicked birch flailed its fiery branches, inadvertently fanning the blaze. It twisted towards Talfryn with frightened eyes that reminded him of Lily and let out a tortured wail.

Instinctively, Talfryn raised his arms to the stars. *"Hydor!"*

Clouds stirred like a celestial soup. Talfryn sensed a thickening swell the sky.

"Luminous, are you recovered enough to take flight?"

The winged horse bobbed her head.

"Then alight and hasten the storm towards the fire."

With a *whap, whap, whap, whap, whap* Luminous galloped into the clouds.

The birch spirit shrieked. Talfryn doused the tree as best he could with a Gentle Rain spell. Flames hissed as the water extinguished them. One smoldering branch broke from the trunk and clunked to the ground.

"More water is coming," Talfryn promised. He ran to the stone cottage just beyond the birch. Fire consumed the thatch roof. In front of the burning home, Vendel hunched over a man's body.

"Dead," Vendel proclaimed. "Not by fire. Axe blow. Did you see anyone?"

Talfryn shook his head, then abruptly turned away. He swiped at his eyes, tearing from the heat and smoke. "What's taking so long? *Hydor!*" he commanded again.

A stifled cry sounded from within the burning house.

"Someone is in there!"

"Ye can't—" Vendel's words were lost to thunder.

Flames danced hungrily through the hole as Talfryn ran towards the opening. Once more, he cast a Gentle Rain spell.

Vendel caught up with him and pushed him aside. "I'll go first, in case there is Cold Iron."

Dodging fiery thatch dropping from the roof, Talfryn followed. Flames devoured a box bed built against the wall, and the sickening smell of scorched animal pelts filled the room. Talfryn coughed as the foul smoke seared his nose and throat. The blaze engulfed an overturned bench and loom. Voracious flames consumed wooden bowls and cups that had been knocked from a shelf, and burning objects toppled to the tamped earth floor.

"Stand clear!" Vendel threw an axe, a sword, and a helmet out the door. "I don't see anyone!"

Sweat matted Talfryn's clothes to his skin. He strained to hear over the fire's roar. A slight movement on the second bed caught his eye. After dodging a length of blazing thatch, he snatched back the pelts.

A small child gasped for air, then shrunk back.

Talfryn's heart jumped in his chest. He reached out to gather up the girl, but she screamed and pounded his arms with her fists.

"I'm trying to save you!"

Thick black smoke curled around the girl's bright red face. Choking, coughing, she still fought.

Vendel reached his side. "She believes ye are one of the men who set fire to the cottage!"

"By the gods!" Talfryn pulled a pinch of fairy dust from his pocket and sprinkled it over the human. Abruptly, the girl quieted and lowered her fists.

"Make haste!" Vendel yelled from the burning doorway after dashing ahead.

Talfryn carried the little blonde outside as the steady rain he had called with the Hydor Spell finally began to fall. Raindrops hissed against the fire-hot weapons Vendel had thrown out the door. He hurried past the strewn Cold Iron, then set the child down, far from the dead body and with her back to it. Talfryn fumbled with his brooch, its gold still hot from the intense temperature inside the house. Finally managing the clasp, he removed the brooch, then his fairy cape. With a soggy snap, he tented the mantle over the young girl.

A raven perched in a nearby tree screeched an aggrieved *kraaak*, then took flight.

"The rain won't last long, but this will keep you dry 'til it stops."

As the girl huddled under the cloak, red-faced and still coughing, Talfryn summoned a Recover spell. Magic tingled and chilled his veins, twitched down his arm and out his fingers, then fell like stardust on the child. Talfryn's shoulders slumped, and his arms hung like broken tree branches. The girl's breathing calmed. She reached for his serpent bracelet and touched its glowing red eyes. *Has Glimmer's protection ended?* Dread burrowed into Talfryn's stomach.

Vendel frowned down at the adder bracelet. "There must have been more Cold Iron in the house."

"I didn't see any."

"Could it be in the ground?" The dwarf's puzzled glance swept all within the blaze's glow.

"Mab save me if it is." Already his flesh burned beneath the bracelet. Talfryn's head buzzed. "My whole body is zinging." Life returned to his arms.

"How odd." The dwarf studied him in the firelight. "Ye don't appear ill." He opened his blistered hands to the welcome rain.

"I could fight an entire fleet of Norsemen." Talfryn summoned another Recover spell and sprinkled it on Vendel's burns. "Do you think I'm somehow becoming more human?"

Vendel flexed his newly healed hands. "Ye just demonstrated courage and strength like yer mother. Mayhap those qualities are helping ye resist the Cold Iron."

"Fine, as long as I don't lose my father's magic."

"Mayhap ye are becoming the best of both yer parents. Let's hope it doesn't wear off."

Talfryn stared down at the child. "She can't be more than five or six summers. Where is the rest of her family?"

"Probably taken. Or..." Vendel glanced at the murdered man's body. "The livestock are gone too."

"We didn't pass anyone." Talfryn glanced back. "So, the men who did this must still be ahead of us."

"Which means—"

Talfryn's stomach twisted. "They are between us and the Callanish Stones."

Chapter Thirty-Six
Lineage

The Blackwood Stronghold

Cold Iron. From her seat at the high table, Fenella studied the broadswords hanging on the walls of the Great Hall. *Symbols of Blackwood power and prowess. They canna be heavier than Highland swords. I could lift one.*

Worry knotted her stomach. *Time is against me. Edward is bound to discover I carry the taint.*

Indeed, Blackwood's eyes were dark with unasked questions, but mostly he appeared relieved to the bone to have her beside him. He sat so close, heat streamed from him, warming her.

"Lady Fenella—" Afore Edward could continue his thought, his father again interrupted him, forcing him to turn away from her and answer more of his father's relentless interrogation.

Would Argonshire nae let him eat? Both men kept their voices low. She squirmed in her chair, wishing she could hear their words.

"Nay appetite?" Connell asked, seated on her other side.

Fenella shook her head. The weight of her mother's words stacked like stones upon her chest. *Forgive Gran…*

He broke off a chunk of bread and handed it to her. "While you readied yourself, I spoke briefly with his lordship regarding your wedding." Her uncle leaned close and whispered, "Your betrothed appeared nay more anxious than you to meet with his father and went directly to his chamber."

Fenella's brows ticked upward.

"I will represent the clan at the wedding. Your Uncle Dunstan will remain behind to guard the thanedom. Your Uncle Artair—"

"Aye?"

"Artair plans to be nearby."

Fenella inclined her head slightly and fingered her goblet. "And Sorley?"

"Like your gran, he was nae invited." Connell glanced down the table, then lowered his voice further. "He will wait in the village with a scattering of allies. Some Dunstan's, some Artair's, and anyone who can slip unnoticed from Castle Thorburn."

Her heart swelled, realizing what a protective presence her uncles had been in her life. But a fresh worry rose. She leaned in close, feigning interest in their shared trencher, and spoke so only her uncle could hear. "The fewer Roses and Thorburns present, the easier to murder me afterwards."

Connell paused while stabbing his oat-stuffed pike.

"His lordship needs a Thorburn to witness my nuptials," she whispered, her head close to his, "then leave so he can do the deed." She drew back and fingered her goblet.

Connell slowly chewed the fish. He angled his head towards Edward, then raised his brows.

Fenella shook her head.

Edward, still being questioned by his father, brushed his fingertips across her thigh. Her insides fluttered. *Did he hear our conversation?* She squeezed his hand, then released it, nae wishing to draw Argonshire's attention. Edward swept his hand back to his lap afore reaching for his ale.

Connell whispered in her ear. "I will send word to the village; warn those who can shift to prepare."

She drew back and stared wide-eyed at her uncle.

One corner of his mouth lifted. "Scholar. Secret keeper. Clan first—"

"Always." *How long has he known?* On impulse, she leaned close and whispered, "Tell Argonshire that Gran is to be invited. I wish her presence."

Connell widened his eyes in evident surprise, then said, "I will attend to the matter."

"I thank you." Fenella slid her cup towards her. As she brought the vessel to her lips, she noticed flames reflected in the amber liquid. Her heart jolted.

Flames?

Fenella stared at the candles on the table, then back at the cup. She twisted to view the rushlights in iron holders on the wall behind her. No matter how she angled the cup, the flames remained. Her mind ground to a halt like a clogged mill wheel.

She cast a furtive, darting glance around the hall. A few men averted their eyes and turned to their companions. Most were intent on their trenchers and ale.

With a slow, deliberate motion, Fenella lowered the goblet to the table.

"Niece?"

"I am a little dizzy, Uncle," she said truthfully.

"Must be a consequence of the whirlwind," he whispered. "Try another bite of bread."

"Aye." She broke off a piece of the dense crusty bread and placed it in her mouth. It tasted like dry leaves.

Connell picked up her cup and eyed its contents. "You still have some ale left. Have a sip. 'Twill calm your nerves." His eyes were guileless.

She took the cup from him and sneaked a peek. The fire had disappeared, but in its place, she spied Talfryn, Vendel, a wee lass, and a shining winged horse.

Fenella's scalp tingled. The room spun, and the cup tumbled from her hand. Ale fanned across the tablecloth in an amber tide as she listed to the side like a ship in a gale. She clutched Connell's arm. Fleetingly, his warmth issued through his linen sleeve.

Vague rustlings rushed her senses, and from seemingly faraway, Blackwood called her name as the world went dark.

****

Fenella did nae fully wake when she became aware of fine quilts and heavy furs pressing her into a mattress. The calming scent of freshly cut lavender prodded her senses. Voices came and went, Edward's soft and worried, Uncle Connell's alarmed and accusatory, then the earl's angry and defensive.

Nurse, the woman who had so poorly cared for her when she had taken fever after being branded and stolen, shooed them away. The chamber quieted.

In the dark stillness, images flashed across Fenella's mental landscape. Women, some old, some young, some timeless, appeared one after the other in rapid succession. Nae stilted portraits, but moving, breathing black-and-white images. Some glared. Others moved their lips as if they were speaking. Fenella wished she could hear them. They all appeared oddly familiar, though she was almost certain she had never met a single one. Had she seen them in the village or in a crowd? She did nae think so. But Fenella sensed each image somehow led her further back in time.

Then in her twilight state, she glimpsed a woman she did recognize–the shape-shifter who had gifted her the fairy horse. Her awareness heightened, she watched with intense concentration as the shifter faded away and a new image appeared. Only the woman's profile was visible. *Could she be the sorceress who had lured Sorley to Fairy, certain in her ruse I would follow?* The image turned to look directly at her.

Fenella's eyes flew open. *The Lady of the Lake!*

Chapter Thirty-Seven
Fork in the Road

The Isle of Lewis

Which name did I hear earlier? Magnus Barefeet?

Talfryn lifted the human child, still docile from the fairy dust, and placed her in front of Vendel, who was already perched upon Luminous's back. He watched the girl finger the string of tiny crystal bells laced around the winged horse's neck. Only once did she glance back at the smoky stone shieling that had been her home. Despite the dusting of magic, tears brimmed her eyes.

A hollow yearning in Talfryn's heart, one left by his mother's absence throughout his life, ached anew. With a sniff, he glanced up and noticed the patter of rain trailed off as the Hydor spell ran its course.

Nay. Fear crackled Talfryn's nerves. He listened for voices carried by the breeze. *Magnus Barelegs.*

"They can't be too far ahead of us," Vendel said, appearing to read his mind.

"Most assuredly." Talfryn flung a pinch of his father's magic granules, conjuring a fresh glittering path that led to the edge of a forest, then forked, with one trail disappearing into the woods, the other on to parts unknown. "I wish I had an elven sword."

"Ye have yer magic."

"And they have Cold Iron. I'd say the advantage is theirs."

"Not necessarily. We know about them, but they don't know about us."

"True." Talfryn felt a little bolder as he set out along the path next to Luminous. "Do you think my mother realizes war bands are headed her way?"

"The Disa I knew never lowered her guard. She will be watching for trouble." Vendel tilted his head towards the girl. "What are yer plans for the little one?"

Talfryn regarded the soot-smeared human. "Mayhap she has family elsewhere on the island. We'll safeguard her 'til we can find them, if they've survived, or find her a suitable home." *She has lost both her parents. At least I have always had my father to guide me.* The dull pain in his heart mushroomed at the thought he might fail, and his failure would cost Beacon his life.

Vendel lifted his gaze. "Then we best move quickly. Gloaming is upon us. Once night falls, the light from Beacon's magic dust will reveal us to the Norsemen."

"Aye, nonetheless, we have little choice but to follow it." Afore him, an enormous oak guarded the fork in the path. The tree's gnarled limbs stretched over the divide like arms tensed for a fight.

"Beware." Vendel hefted the blackthorn staff. "A guardian watches."

"I see it."

Luminous whinnied, then retreated a pace.

The fine hairs on the back of Talfryn's neck raised like hackles as he sensed the spirit in the oak assess him. Beyond the guardian, wood nymphs whispered to one another.

"We've been spotted by wood nymphs, as well," Vendel murmured.

Talfryn nodded, weighing his next move. "There must be fairies in these woods. They'll tell us how to find the Callanish Stones."

"Might be a goblin wood," Vendel countered. "We could be trapped, outnumbered."

Sniffing the air for whiffs of the damp dog scent of goblins, Talfryn thought, for a heart-stopping moment, he detected Cold Iron.

What choice would Father make? "We'll trust the magic." After pulling the child from Luminous's back, he whisked his cloak off the girl and fastened it over his shoulders. Then he conjured a scarlet mantle like Fenella's and bundled it around the little blonde. The young human stood mutely beside him, her left hand clasping his leggings.

Vendel dismounted.

Forehead to forehead, Talfryn pressed against the winged horse. As he stroked the magical creature's cheek and neck, he whispered, "How did you know to follow us?"

After a pause, Vendel answered for the horse. "Lily tracked her down to the Borders, where two trolls had caught Luminous unawares and captured her with a bespelled golden rope."

"Nay! The Borders? You could have been killed!"

Luminous stared at Vendel as if sending him her thoughts.

"She said she just touched down for a moment to drink from a stream. The trolls boasted how someone high in the Blessed Court would pay them well to make sure you never saw Luminous again."

Talfryn recoiled, then stared into the winged horse's eye. "Why would Mab or any of The Six want to separate us?" As he said the words, the answer came to him. "Because I favor you over equistragons."

Luminous nodded.

"Lily rescued her, then beseeched Luminous to trail you and help when needed." Vendel jabbed the blackthorn in the air. "I knew this was Lily's doing!"

"Why didn't you say something sooner?" Talfryn demanded.

"I thought 'twas obvious."

It was obvious, he realized. *Only Lily would take measure of my peril, then risk her own safety to aid and safeguard me.* Lily had befriended him when other Trooping Fairies had snubbed him. She never refused a challenge or adventure. Even when some nobles taunted her for having ears they thought too pointed, too elfish, Lily held her chin high. His heart expanded as he remembered how Lily had dared show her burns after Merrybelle's murder.

Talfryn stroked the steed's neck. "I am grateful for your aid, Luminous. We would have been dead by now without you."

Luminous nodded.

"I am in your debt," he whispered in her ear, then he flashed on her plight should she remain within Mab's reach. "But now you must fly far from me, to somewhere secret and safe. Mayhap Merlin will know of a place. Seek him in Avalon or in his Enchanted Forest. May Fate bring us together again in a fairer future so I may repay what is owed."

The winged horse nuzzled his shoulder. Small vaporous clouds huffed out of her velvety nose.

"Farewell, my friend."

Luminous bowed her elegant head, then took to the air.

Beside him, the young girl, wide-eyed and slack-jawed, watched the powerful, enchanting creature fly into the night sky.

Talfryn clasped the child's hand, turned to face the forest, then decisively strode forward.

With a loud creak, the oak swung one of its long, thick branches downward, grazing Talfryn's chest as it blocked his way. An ancient face with sunken slanted eyes, long nose, and mustache-framed mouth, appeared near the base of the oak. "Do not take another step!" it ordered in a voice rasped by time, wind, and smoke.

Talfryn swept the girl behind him. "We wish to follow the illuminated path, Ancient One."

"Do you, now?" The oak spirit slowly lifted a root, tearing mossy clumps of soil from the earth, to scratch its beard. "Why should I let a firestarter pass?"

"I am not a firestarter." Talfryn glanced down at his Norse tunic and breeches. "This is but a fairy glamour."

"Why would a fairy dress like a killer?"

"'Tis a necessary disguise. While searching for someone in the Norse lands, a Norsewoman, we were told she might be found at the Callanish Stones."

"And why do you seek this woman?" asked the tree spirit.

Talfryn glanced at Vendel. "She stole a vial that belongs to Mab, the Queen of Fairy. Our dear queen wants it back."

"Thieves and killers," the tree spirit muttered.

"May we pass?" Talfryn itched beneath his coarse wool clothing.

The oak's gaze scraped over him. "Prove you are fairy. Remove your glamour."

"This lad hails from the Noble House of Storms." Vendel thumped the blackthorn staff on the ground. "'Tis one of the oldest Trooping Fairy families. The lad doesn't have to prove anything to ye. Let us pass!"

With an abrupt brandishing of its branches, launching an unkindness of ravens hidden in its limbs, the ancient oak bellowed, "*This* is the last magical forest, the only sacred wildwood on the island torchbearers haven't set ablaze. None of you may proceed 'til he removes his glamour!"

Behind Talfryn, the small human child sobbed and pressed against his leg. Unsettled by the girl's distress, Talfryn flicked his hand, and a miner's gray tunic and breeches replaced Vendel's Norse garb.

"See!" Vendel exclaimed.

"Now you." The oak glared at Talfryn.

Talfryn cast a sideways glance at Vendel. Mayhap Glimmer's spell had worn off. He concentrated. A faint wave of magic rippled from his head to his boots. Cumbersome muscles weighed heavily on his formerly lithe frame, and his once short spiky hair stuck to his sweaty nape. At least his clothes had changed back.

The oak leaned forward with considerable creaking and rustling.

Talfryn held his breath.

"Not good enough," the oak declared.

"Heed this human." Talfryn stepped sideways so the oak spirit could see the child. "We rescued her from a burning shieling. That proves we aren't the ones setting the fires."

"Little Signy." The oak softened his tone as he creaked farther forward and squinted. "She reeks of smoke, and her skin and clothes are sooty. Whether you rescued her or stole her cannot be proved. Where are her parents?"

Talfryn pressed his hands over the child's ears. "Her father is dead. We presume her mother was taken."

"Have their gods forsaken them?" The oak flailed its branches. "May an axe fell those responsible! At least forest pixies intercepted the Norsemen near our northern border."

"Torchbearers?" Vendel asked.

"Aye. Pixie scouts diverted the band. They'll push them towards the shore."

"And if there are others?" Talfryn asked.

The oak swayed its branches in a forlorn gesture.

"Then Mab speed to your pixies."

"I thank you." The oak studied him. "No Norseman has your fairy eyes. And none of you carry Cold Iron or torches. Therefore, I will grant you entry into the wildwood. It is the shortest route to the standing stones...if you can survive what waits within."

"We thank you for granting us passage." Whatever fantasies remained about his mother's people drained away. "Forgive us, Ancient Guardian, for disturbing you on this night of terrible fire.

With a laborious lift of its limbs, the tree twisted, allowing them passage.

Talfryn bowed to the oak, then led Signy forward.

Once they were past the guardian, Vendel warned, "The forest will be the last refuge for the island's magical folk. Trolls. Goblins. Outcasts—"

"Nay need to frighten Signy further. Just watch for signs of danger."

"Such as?" Vendel asked.

"Worry when mushrooms start glowing, roots crackle, or trees release foul, deadly scents through their leaves."

"Glow. Crackle. Stink. Lovely. What about will-o'-the-wisp lights?"

"Only in bogs." Talfryn squared his too wide shoulders, picked up Signy, then ducked into the thick, forbidding overgrowth.

Chapter Thirty-Eight
The Last Forest

Trees huddled together, blocking the sky, adding to the deep darkness. Talfryn sensed dozens of eyes assessing him, determining if he were fay or firestarter, while a queer quiet hovered, as though the forest held its breath. Silent. Still. Wary.

When Signy whimpered, Talfryn squeezed her hand. Anywhere else, any other time, he would have conjured orbs to light the way. But not now when brittle tensions permeated the forest, and the sour stench of fear tainted the air.

"I will lead," Vendel said. "'Tis lighter here than in the crystal mines."

Talfryn nodded, then glanced back as three tree spirits shifted together, blocking any chance to retreat. Unease scrabbled down his spine. Trenching his brow, he peered into the darkness. The enchanted path extended straight ahead, casting a faint orange glow on the tree trunks.

"I see birch."

"Aye," Vendel whispered. "Ash and hazel too."

"Nay more oak. Any rowan?"

"None that I can see. Ye should be safe," Vendel said.

Taking a hesitant step forward, Talfryn caught a whiff of damp dog. *Goblin.* His pulse zigzagged when he saw a small goblin, its hairy body hunched over a red lantern, flash amber cat eyes at them, then scurry across the path and disappear into the woods.

Vendel exchanged a fleeting yet knowing glance with him.

"We're going to play a game," Talfryn whispered in Signy's ear. "You climb on my back, and I will be your horse."

Signy shook her head, then fisted his leggings and pressed her forehead against his leg. Talfryn squatted and hummed a fairy tune in her ear. She relaxed and loosened her grip. The fairy song swelled, lifting her in the air. He cupped his hands into stirrups and turned so she could climb on his back.

"Not so tight," he rasped, adjusting Signy's death hold on his throat. "Wrap your legs around my waist."

After the girl had settled, a moss maiden at the base of a nearby alder burst into nervous laughter and was quickly silenced by a wood nymph.

Moss maidens, goblins, and tree spirits, all in this one place, their last refuge.

"They'll be frightened and desperate," Vendel said, apparently reading his mind again.

"Make haste, and only use defensive magic." Talfryn tilted his head towards the blackthorn.

Vendel glanced at Signy, then nodded.

Talfryn dug into his pocket and scooped up the remaining pixie dust.

A horrible wail tore through the trees. Signy's slim arms tightened around Talfryn's neck.

"Banshee." Vendel wielded the blackthorn staff in front of him like a truncheon. "Bad luck to hear a banshee, much worse to cross its path. Could mean someone's going to die."

Gritting his teeth, Talfryn crept forward, ducking low branches bearing tentacles of moss. *Hope we only hear it then.*

A nymph in a young birch reached out her long-fingered hand and snagged Talfryn's cloak.

Startled, Talfryn looked over his shoulder at her, relieved she was only a nymph and not the banshee.

"Are you going to set us afire?" Her voice trembled.

"Nay," Talfryn said. "I brought rain to douse the fires."

"Rain?" The nymph appeared confused.

"Aye, a Hydor spell." Talfryn glanced up at the thick tree canopy. "Mayhap it didn't reach you."

"Could you send more?" the nymph asked.

"I could douse you with Gentle Rain. It isn't much, but it is wet."

"Oh, please do!" The nymph bent forward.

Talfryn raised his hand, waited for the familiar tickle in his palm, then sprayed the birch's upper branches.

"Much better!" the nymph said in a shivery voice.

High above his head, other birch nymphs conversed in quick, feminine whispers. Higher still, ash argued in deep, masculine voices. Their angry murmurs reached Talfryn's ears. The surrounding birches released a warning malodor.

Beneath Talfryn's feet, the earth quaked. Then it stilled. Something scraped. Again, the ground quaked.

The young nymph's eyes bulged. "A troll approaches!"

"'Tis the henkie," a nearby hazel said in a quivering voice.

"A henk—" Talfryn stopped mid word. An emaciated troll, nearly as tall as a full-grown hazel, limped onto the sparkling path. Moss and twigs ensnared its greasy hair, and tears streaked its scarred face.

All the tree chatter stopped. Fungi growing at the base of the trees began to glow a soft blue warning.

Vendel ducked behind the hazel. Talfryn followed, then groped in his pocket for more pixie dust and grabbed the last few specks.

The henkie sobbed, a desperate, earsplitting sound that evoked images of caged banshees and trapped dragons. It flailed its overly long arms.

"What's wrong with it?" Talfryn asked.

"Poor one's offspring is lost in the woods," the hazel explained.

A leafy-faced moss maiden, smaller than Signy, emerged from the lichen near the exposed roots of the tree. "I know where 'tis," she said. "'Tis asleep at the base of a birch, a good way down the path."

"Are you certain?" Talfryn asked.

With an anxious glance at the henkie, the moss maiden inclined her leafy head.

"The creature is wounded" Vendel warned, "and dangerous."

Talfryn studied the henkie. "More desperate than dangerous, and in need of a Restore spell."

"A Restore spell would leave you weak and vulnerable."

"Aye, but we'll be able to cross the woods with great haste and unharmed if we travel with the troll. If we help him; he will help us." *Besides, he needs our aid.*

"Ye'll help yerself be killed. Then what will happen to Signy and me?"

"Stay with Vendel." Talfryn said as he kneeled so Signy could slip off his back. She stared unblinking at the moss maiden.

"Remain behind the henkie," Talfryn told Vendel, "lest the creature rampages."

Grim-faced, Vendel nodded.

The Halfling rose, then gave the dwarf a reassuring pat on the shoulder. Muscles taut, Talfryn emerged from behind the hazel and approached the weeping henkie. The troll's mouth hung open, and in the dim light, Talfryn noted bloody scratches on its face. *Poor creature has probably crashed into branches while flailing about.* With his hands outstretched to show he was unarmed, Talfryn approached the towering troll.

"The moss maiden will lead you to your child."

Fresh tears spilled from the henkie's swollen eyes as it emitted another earsplitting cry.

"But you have to move quickly." Seeing the henkie leaned on its good leg, one shoulder lower than the other, Talfryn hesitated. "About your leg…" Casting a Restore spell *would* severely weaken him, and if the troll grew violent, he'd be unable to protect himself, much less Signy and Vendel. *The dwarf has the blackthorn. He and Signy will be safe.*

Talfryn raised his right hand towards the wound. Magic tingled along his veins. It sparked under his skin, hummed around his knuckles, then shot out his fingertips. Glittering sky-blue ethereal filaments wrapped around the troll from his cloth-covered waist to his boots fashioned from rags and deerskin.

"HMMPH?" The henkie straightened, then bent its bowed legs a couple of times, testing their strength. A flash of relief flitted in its dark eyes, then died.

The familiar drain of vigor from the Restore spell sagged Talfryn's shoulders. His arms hung like deadwood. *Mab save us if we are attacked.*

Her lichen-covered feet and ankles illuminated by Beacon's magic dust, the fern-haired moss maiden scurried ahead.

Talfryn trudged after her, ducking low branches laden with moss beards. "This way," he urged the henkie.

Vendel and Signy trailed the lumbering creature.

Soon, apparently fully healed, the henkie gathered speed, taking long, quick strides. Talfryn dove out of its way. The moss maiden glanced back, and seeing the troll bearing down on her, quickened her pace.

Jumpy with urgency and with one eye on the henkie, Talfryn waited for Vendel and Signy. They reached him as Talfryn's arms prickled back to life. "Back on the horse!" he declared as he whisked up Signy. She scrambled onto his back like a blond beetle.

"I will keep you in sight!" Talfryn promised Vendel. Trees flanking the path blurred as he performed a running leap, then another, jolting Signy each time. Vendel's heavy breathing and quick footfalls faded as he fell behind.

With occasional pauses, Talfryn kept the moss maiden and henkie in view ahead of him while giving Vendel a chance to catch up.

At last, the moss maiden scampered to a halt in front of a birch alongside the path. Her breath raspy, she pointed, then scurried off.

Moving with renewed speed, Talfryn caught up with the henkie.

A tree spirit with a shriveled prune face materialized from the birch, then arched one thin brow. "You should have asked the trees instead of weeping about," it scolded the henkie.

Notes of relief and gratitude spiraled through the henkie's cry as the creature sank to its knees and gently patted a large, leaf-covered lump at the base of the birch. Fungi huddled along the forest floor ceased glowing when the mound stirred, poked out a pale arm, and sighed contentedly at its father.

Vendel huffed up to Talfryn and leaned on his wooden staff. Once he caught his breath, he wheezed, "Ye were right. We crossed a lot of ground and not a soul bothered us."

The henkie uttered a garbled sound, gratitude glistening in its sunken, moist eyes.

"We are happy you found your baby," Talfryn told the troll as he returned the pixie dust to his pocket. "Good fortune and farewell." Then he turned to Vendel. "Let's see how much deeper these woods go."

They walked as quickly as Vendel's short legs and the dark forest allowed. Talfryn kept his ears pricked for the snap of a twig or the whistle of an elf arrow. He had not forgotten the banshee's wail they had heard upon entering the forest.

Soon, they reached a spot where ash trees crowded the path and decay thickened the air. A ground mist threaded between the trunks, and though the cloud only reached Talfryn's knees, cold dampness sheened his face.

Deep in the forest, a ghostly bark split the silence.

Vendel glanced over his shoulder. "What was that?"

Talfryn peered into the darkness. "A Cu Sith," he said in low voice.

"Coo shee?" Vendel asked.

"Aye… a dog the size of a bull. They guard fairy mounds. Far more dangerous than Arwan's hounds." He hoisted Signy higher on his back. "Nay human has ever survived hearing a Cu Sith bark thrice."

"Do you think it caught our scent?"

"Let's hope not. The Cu Sith's bark can carry for leagues. Nay telling how close the beast is." Nervous but determined, he nudged Vendel, and they inched forward. A brittle branch snapped beneath Talfryn's boots. That's when he noticed the sudden stark silence blanketing the forest. He dared not move.

A nearby nymph faded into a tree trunk.

"I don't hear anything," Vendel whispered.

"They glide silently, but always in a straight line."

Vendel released a long breath. "Then we're safe. The trees are too dense. Only this magic path allows us passage."

Despite Signy's warm breath on his neck, Talfryn felt the fine hairs on his nape rise. "What if the Cu Sith is on this path?"

They both whirled and stared. Seeing nothing, by silent agreement, they turned back and broke into a run.

"Are we being pixie-led?" Talfryn whispered when they stopped to catch their breath. "This forest seems everlasting."

Afore Vendel could answer, a second spectral bark, louder and closer than the first, exploded behind them.

Signy screamed.

Tree roots crackled.

"Climb to safety!" Talfryn hoisted Vendel into the nearest tree, then handed up Signy. "Don't let her fall."

Vendel clasped the girl's wrists and pulled her up beside him on a hefty branch. "And ye?" he asked Talfryn.

Suddenly dizzy, Talfryn staggered back.

"That was the second bark. Climb up!"

"I can't. I…am…unwell..."

"'Tis a rowan! Move away!"

"Rowan?" Talfryn pressed his palms against his throbbing temples. His thoughts swam through murky memories. Hadn't Fenella said something about rowan?

"Rowan wards off fairies. Move away from it afore the third—"

Terrifyingly close, the Cu Sith bayed.

"Move!" Vendel yelled.

Talfryn stumbled backwards. Shapes blurred afore him. He thought he spotted a hazel a few trees up the path and staggered towards it. "Stay up there, nay matter what happens," he ordered Vendel. "I'll draw it away from you."

"Ye are half human!" Vendel protested.

"Shh!" The hazel was in arm's length; he just had to put one foot in front of the other. He fluttered his eyelids, but his vision remained unsettled. *I must move off the path, afore it sees—*

Talfryn sensed the large presence prowling behind him. Then hot, moist breath dampened his back as a deep, deadly growl shuddered his body.

Chapter Thirty-Nine
Seeing

The Blackwood Stronghold

"Viviane!" Fenella cried.

The ceramic pitcher slipped through Nurse's fingers and crashed to the stone floor.

Fenella flinched.

"Merciful saints, Thaness! You gave me a start waking like that. Who pray tell is Viviane? Nay soul here answers to such name." Nurse had halted halfway between the door and the four-poster bed. Her gaze now swiveled between the two. "I'll send Derry in to clean the spill. And his lordship will want to know you have wakened."

As if a fairy hound nipped at her heels, Nurse hurried out of the room and pulled the thick wooden door closed with a click. Fenella sat up in bed and stared at, without really seeing, the massive door. Her gaze drifted to the floor where rosemary water pooled like a shallow lake among red-pottery-shard islands.

Viviane.

The skin on Fenella's scalp crawled. She eyed the puddle as she threw back the covers, scooted out of bed, and took three dizzying steps across the cold stone floor. Kneeling beside the shards, she gathered her hair into a loose knot at her nape. When she leaned forward, her freckled face, unexpectedly pale, stared back at her. "Take heed, water. Show me Talfryn, the Halfling."

A small green face with close-set eyes appeared in the water.

"Och!" Fenella drew back.

"What are we looking at?" Puffball asked.

"Not *at, for.*"

Puffball's ears twitched, and she tilted her head to one side. "Were you ill? Your Mr. Blackwood carried you up the tower stairs. His father forced him to leave."

Edward, with his injured leg, carried me? Her heart swelled, then grew heavy. "I am nay ill. Least, I doona think so. I saw Talfryn and Vendel reflected in my ale cup. Then I must have fainted. Afore I woke, I dreamt of the shape-shifting woman who gave me the fairy horse and—"

"Fairy horse? Was it Viviane's, the Lady of the Lake's?"

"Nay, though the woman and horse made me think of Avalon. An old woman stopped us along the road. When I glanced back, she appeared much younger and had hair like mine."

Puffball scratched beneath her skullcap. "Viviane has hair like yours, and a white fairy horse, but I've never heard tales of her shape-shifting. Sounds more like Morgan le Fay. She's a shifter."

Fenella clasped her hand over her mouth. "Fate's wings! King Arthur's half sister? Do you know her?"

"I know *of* her. Morgan le Fay, the water fairy who married a human king. Did you dream of anyone else?"

"Aye. I saw Viviane." She glanced at the puddle.

"And now you're drawn to water?"

Fenella nodded.

"Hmm." Puffball crouched on her spindly legs and plucked a few shards and rosemary twigs from the water. After she had placed them aside, she asked, "See anything now?"

Leaning forward, Fenella peered into the water. "Aye! Talfryn is carrying the wee lass! But I doona see the winged horse anymore." She picked the broken jug handle from the puddle. The water rippled, blurring the images.

"Do you see Disa?"

As the water stilled, Fenella watched a dark forest take shape. She could just make out a towering form emerging onto a path lit by glowing orange specks. Engrossed, she did nae hear the chamber door creak open.

"Thaness?"

Fenella jumped back. Her hair tumbled from its knot.

"My lady!" Derry propped a knobby-handled broom beside the door and placed a green-glazed jug upon the bench. "The earl would dismiss Nurse and me if he found you picking up broken—" Derry spotted Puffball. Surprise, shock, and delight played in turn across her heart-shaped face. The stave bucket hooked over her wrist clattered to the floor.

Then, unmistakably, Derry's dull blue eyes brightened and tilted.

Fenella rose and padded over to her. "Derry?"

The maid reddened. Her hair, usually a lackluster mousey shade, glinted as if burnished.

Fenella closed the door. "Does fairy blood flow within your veins?"

Derry raised a trembling hand to her throat. "What a daft thought!" She gave a nervous laugh that ended in a strangled snort.

Fenella stared at her.

As the reddish hue deepened on Derry's cheeks, she gulped. Then her expression grew serious. She lowered her voice. "The earl ordered Nurse to report anything strange you say or do." Her glance flicked to Puffball. "But I doubt he would deny your wedding now if you were the devil's spawn."

"I suspected as much." She snatched her scarlet Thorburn mantle from a peg and wrapped it like a mother's hug around her shoulders.

"The earl would have Mr. Blackwood marry you, become Thane, and then…"

"I would meet with an unfortunate accident?"

"I fear so." Derry stooped to pick up the bucket. "Nurse is on her way to see the earl now. And I willna be long for the castle either if he suspects anything unusual in my lineage."

"Then we will nae tell him. Though how you manage around the Cold Iron is a mystery."

"Could work if she doesn't have too much fairy blood," Puffball said.

Derry cast her eyes downward and spoke in a soft tone. "You cleared the Cold Iron and rowan from this chamber, save your dirk. I am much obliged." The maid kept her gaze rooted to the floor and approached the spill. She stooped and began gathering the broken shards. Pottery pieces clattered into the bucket.

Fenella joined her. "Derry, find my uncle, Mr. Thorburn. Tell him he must invent a distant cousin named Viviane."

"Aye, Thaness." Derry scooped up the last shard, tossed it in the bucket, and stood. "Horse tails and wolfsbane! I forgot to bring a cloth to sop the spill. I will fetch one, then sweep the floor for splinters. You should hide your little friend, and your emerald cloak. From what I've overheard, nae that I listen at doors, the earl is already suspicious of the mantle."

"Och!" Fenella grabbed the garment from the peg and rolled it into as small a bundle as possible. "Puffball, hide!" she said as she stuffed the shimmering green cloak behind the four-poster.

An unbidden memory flashed through her mind as she handled the thick, but curiously light, cloak. A stark, clear vision of how the beautiful material had *called* to her. Viviane's emissary, Glynis, had taken her to the tailor shop to have a mantle made. The cloth had been stacked on a high shelf, and the tailor had nay memory of purchasing the fabric. 'Twas nae long afore she and Talfryn had parted…he back to Fairy, she to Edward.

As thoughts of Edward forced images of the tailor shop from her mind, Fenella gave one last resolute shove to hide the covering from view, then followed Derry to the door.

Derry disappeared down the tower stairs. As Fenella closed the door behind the maid, she noticed with dismay Calidore was nae at his post.

"Can you see me?" Puffball peeped from the narrow crack between the four-poster and the clothes chest.

"Nay," Fenella said absently. She paced, unsure what to do next. Footsteps sounded in the hall, accompanied by the low murmur of conversation. Pressing her ear to the door, she held her breath and listened.

"We will tell her uncle she is unwell," the earl said in his I'll-brook-nay-argument voice. "After what happened at supper, anyone would believe it."

"But surely, Father, Thorburn will ask to see her," Edward argued.

"We will tell him she is asleep. And well she will be once Nurse returns with the sleeping potion. Thorburn will see his niece when she marries you."

"But she has nae done anything wrong. You canna—"

"I remind you, Edward, whose house this is."

Fenella held her breath.

"Aye, Father. Please allow me to see her."

"Nay, 'twould be unwise." The footsteps stopped. "I wish Gavin were here," the earl muttered.

"I wish Hamish were here," Edward retorted.

A slap, hard and loud, sounded through the door. Then a thump followed, as though the victim had been knocked off-balance and had thudded back against the tower wall. Fenella clapped her hand over her mouth to stifle a gasp.

"Do nae speak of your brother!" Fury and contempt edged the earl's words. "Hamish is with the monks. We shall nae see him again."

Fenella retreated, expecting the door to be pushed open. Instead, she heard the unmistakable click of a key in the lock.

"Edward?" She grabbed the iron knob. It refused to turn. "Edward!"

Argonshire lowered his voice. "Nurse will return in a moment. The thaness will be fast asleep soon enough."

"But—"

"We are through here," Argonshire said, his icy tone echoed along the stairwell as the two men's footfalls softly thudded down the stone steps.

"Forget them." Puffball wiggled out of her hiding place. "Gaze into the water. We must find Talfryn."

Chapter Forty
The Lady of The Forest

Sensing the immense bulk of the beast, Talfryn ever so slowly turned his head to look over his shoulder. Blazing red-orange eyes pierced the darkness. The tips of the fairy dog's fur glowed dark green, revealing its true size.

Talfryn dropped his gaze to the path. *Can I step off more quickly than the Cu Sith can strike? Nay. I fear I'm too weakened.*

Rapid breaths gusted hotly against Talfryn's nape as he slid one foot towards the edge of the path. A growl rumbled up from the Cu Sith's massive chest. Talfryn abruptly halted.

If it were an untamed equistragon, I would not turn my back on it. He willed himself to gradually pivot.

The Cu Sith retracted its upper lip and growled. Its tongue slid across its deadly fangs, flicking foam, and what looked like blood, onto Talfryn's boots.

"Return to your mound!" Talfryn ordered.

With a snarl, the fairy dog curved its plaited tail over its broad back.

"I am a Trooping Fairy from the Noble House of Storms." Talfryn straightened his spine, adding height to his Norse frame. He leaned closer to the beast's gnashing teeth and caught a whiff of death and decay. His stomach churned. "Obey my command!"

The Cu Sith coiled to pounce.

Talfryn tensed to fling a Stray Sod spell. Afore he could cast it, deep elemental magic rippled between him and the Cu Sith. Talfryn's body seized up, joints locking, muscles cramping, preventing him from warding off the enchantment.

The fairy dog's eyes bulged as it, too, became immobile.

Trees surrounding them rustled, bowing and murmuring like flower fairies at Court. Leaves floated gently to the ground as they parted, creating a small clearing. A radiant white stag bearing the Lady of the Forest on its back calmly trod into view.

"Very courageous," the mistress of the woods said. "Especially since the Cu Sith thinks you human, and therefore prey. Are you human? You claim to be a Trooping Fairy, but you appear to be a Norse blaze setter."

After a moment of scrutiny, the Lady sketched a symbol in the air, and Talfryn felt his jaws, head, and neck unlock.

He gulped down his fear.

The Cu Sith glared, motionless, apparently still frozen by the Lady's magic.

Blue damselflies clung to lush ferns that grew in place of hair on the Lady's head. The ferns swayed as she dismounted. Starlight angling into the clearing cast white light on her hazel skin, illuminating fine lines around her pale green-gray eyes. A short-eared owl emerged from the dark forest depths, flying soundlessly above her, then perched high in an ash.

Froth from the Cu Sith's mouth dripped on Talfryn's shoes.

"Declare your intent," the Lady of the Forest demanded, "afore I release the rest of the spell and leave you to your fate."

"Nay fire has been set, nor will be set, by my hand. I am a Halfling," Talfryn admitted. "This Norse glamour was forged by Glimmer, the Royal Sister, and it is loath to leave me."

"I see." The fur collar around the Lady's neck stirred. Two long ears popped into view followed by dark eyes set within a furry, yellowish-brown face. The rabbit's whiskers twitched. Spotting the Cu Sith, it flattened its ears and hid its face in the fur of an adjoining rabbit.

"You say you have set nay fires, yet your hair and skin are smudged with smoke and soot."

"Caused when I ran into a burning shieling to save a young human."

"Where is the child now?"

"Protected."

The Lady's voice surged dangerously. "Nay child of this isle needs protection from me!"

"Nay insult intended, my lady. But there are shape-shifters about. It is difficult to know whom to trust."

"Granted." Appearing calmer, the Lady lifted her gaze and scanned the trees. Talfryn's breath caught when she stared in the direction of Vendel and Signy's hiding place. She paused, and her eyes softened.

"I will mark a symbol in the air," said the mistress of the woods, "which will release the magic holding you and the Cu Sith. If you move with fairy speed afore the beast attacks, your life will be spared."

Talfryn dared not shift his gaze while the Lady raised her arm, pointed her finger, and drew a sigil that shattered the Immobilizing spell. He flung the Stray Sod spell as he leapt to the side.

Snarling, the Cu Sith lunged, its colossal jaws snapping, and bit the air where Talfryn had stood. Stray Sod, reeking of brimstone and fermented poppies, materialized on the path afore it, and the fairy dog's enormous paws landed on the magic plant. The creature bayed, sounding furious and confused, then sat heavily. Talfryn held his breath 'til the Cu Sith recovered enough to stand and stagger forward. When the sparkling trail curved, the beast veered from its orange glow, continuing its straight course 'til it disappeared among the trees.

Talfryn's muscles sagged as he released a long breath. He hoped the Lady of the Forest could not see his legs trembling.

"You have fairy speed and magic. But if you cannot remove your glamour," the Lady said, "how can I be certain you are not a Norse torchbearer? An evil mage mayhap?"

"There are other ways to prove I am a Trooping Fairy," Talfryn said, his eyesight finally clearing, having gained distance from the rowan. "I could create a whirlwind. But the war bands have set fires not far from here, and any flurries I summon would fan the flames."

"Endangering my forest."

"Exactly."

The Lady frowned in the direction the behemoth fairy dog had run. "You ordered it to return to its mound. It cannot. My owl has informed me the Cu Sith's fairy mound has been surrounded by fire. The fay have abandoned the dog to its fate." Her sigh accompanied a momentary slump of her regal bearing.

Unsettled by the news, Talfryn watched the Lady advance towards him, her graceful stride rustling her pine-green gown and its surcoat of leaves. She drew close enough for Talfryn to smell the wild mint on her breath. Wood mice, pygmy shrews, and rabbits formed a circle at her feet. "Why have you entered my forest?"

"I am on an urgent quest to find the Norsewoman, Disa. I believe she may have settled on this island."

"Urgent, and yet you helped the henkie?"

"I can't bear a parent and babe to be lost to one another." *And it kept others away, allowing us to move quickly.*

"Fairies are not known for having soft hearts regarding parents and children. Hence the many changelings in the world."

Talfryn lowered his gaze. "As I said, I am a Halfling."

She inclined her head, then approached the rowan concealing Vendel and Signy. "'Tis safe to come down now."

"Thank ye, my lady," Vendel said.

"Come to me, wee one." The Lady of the Forest reached up and gathered Signy in her arms. One rabbit shifted away from the other as the child nestled against them, then closed her eyes. The Lady rubbed Signy's back, and the girl sighed.

Vendel handed down the blackthorn staff, then waited while the Lady of the Forest carried the magical gift and the girl out of his way. Unencumbered, Vendel sought a lower branch afore he jumped heavily to the ground.

"My lady." He bowed.

"Dwarf," she said kindly. She handed him the rod along with a water flask made of woven leaves.

The Lady turned and walked towards Talfryn, then conjured a second flask and handed it to him. "Refresh yourself."

Unsure if like the fay, the Lady would require a favor in exchange, he hesitated but saw nay expectation reflected in her guileless eyes. "You are most kind. The child seems at ease with you, my lady."

"I was her mother's midwife. Signy's parents, Astrid and Dag, were good people. Sadly, I know of nay kin." Tears brimmed her eyes. "If I thought I would survive these perilous hours, I would safeguard her. But..." She shook her head, then lowered Signy to the ground. The little girl shivered.

"Poor child." The Lady removed a lamb's wool mantle from the white stag's back and added it to the scarlet mantle draped over Signy's shoulders. Then she conjured an intricate brooch made of twigs and vines and fastened the fabric together at Signy's right shoulder with a honed antler pin.

"Please keep her safe," the Lady of the Forest implored.

Talfryn, his chest tight and aching, dropped to one knee. He clasped Signy's hand and gently drew her to him afore rising.

"I have a second but equally vital request of you." The Lady of the Forest angled her hand into a pocket beneath her surcoat. She withdrew a small woven hemp bag and pressed it into Talfryn's hand. "This pouch contains seeds from all the trees in my forest, except the rowan." She leaned forward and whispered in his ear. "See that these seeds are planted where the fairies can protect them, for my wildwood may not last the night."

"You have my vow." Talfryn's gaze lifted to the alders, hazels, ash, and birch trees. A few nymphs clasped their leafy hands. Others lifted their chins skyward, sniffing the air. Their worry and fear descended upon him like a weighty cloak. He stuffed the pouch inside his pocket.

Then the Lady fished beneath a speckled rabbit and withdrew another pouch. The welcome scent of magically roasted chestnuts and warm pine nuts filled the air. "These will give you strength. Eat so you can bring Signy and the seeds to safety." She handed the sack to Talfryn, then plucked one for Vendel, and another for the little girl. "You must not fail in the tasks I have given you."

"I will not fail," Talfryn promised. "But I must beg your leave, dear Lady. Our quest is most urgent."

"Away with you then. Continue along the path 'til you reach a lone willow. Look to the sacred well at its base." Her intense gaze widened to include Vendel. "Be warned. You cannot return the way you came."

"We will find another way. But what of you, my lady? Will you not accompany us to seek refuge?" Talfryn asked.

"I will protect my forest." She signaled the stag. The regal snowy beast ambled towards her. "Do not linger on this island. Its soil is riddled with Cold Iron."

Talfryn glanced down at the glowing serpent bracelet and grew queasy.

"Be your true self, Halfling, and your course will become clear." She mounted and sat sideways upon the stag, then blew Signy a kiss.

"My lady." Vendel stepped forward. "Perchance, have you seen any of my kind about?"

"Dwarfs? Nay. I am sorry."

"I thank you, my lady." Vendel bowed as the stag stole into the woods with the stalwart Lady of the Forest on its back.

Vendel angled his head towards Signy. "She will slow us down."

"We'll manage." Talfryn knelt and formed stirrups with his hands. Once Signy was secure upon his back, he pushed onward beneath the dense tree canopy with only the magical path to light the way. The dwarf trudged behind him.

A high wind stirred the treetops. Smoke whirled into the forest. Through his boots, Talfryn felt thick tendrils beneath the soil reach for one another. Winds grew stronger, knocking branches to the ground. The roots clung to each other, countering gusts that threatened to topple nearby trees.

Live embers and ash will follow. "Make haste!"

The earth undulated, and a tree nymph screamed.

Signy flinched, and Talfryn struggled to keep his balance as he lurched forward.

Vendel stomped on an ember that floated down between two birches, then hastened to keep pace.

Of a sudden, the ground stopped rolling. Pausing to glance about, Talfryn observed aloud, "The wind has died."

"Thank the gods." Vendel pointed the blackthorn staff at a stately tree near the edge of the woods. "Be that the willow?"

Heartbeat spiking, Talfryn ran to the willow, nearly choking when Signy tightened her hold around his neck. The willow's graceful branches swept the forest floor. Spying the well, Talfryn knelt on the damp earth and eased Signy to the ground.

"Now what?" Vendel asked, huffing for air as he reached Talfryn's side.

"Witches scry with water. Mayhap we're meant to do the same."

"Use it like a large Finding Horn?"

"Exactly."

"Rune." Signy plucked a birch chip from the stones ringing the well. She turned it over in her hand and studied the carving on its face. "Frigg. All-Mother."

Chill bumps erupted on Talfryn's forearms. "Father said my mother carried a rune such as this with her everywhere."

"It could have been left by anyone," Vendel countered.

"Only my mother would have left a tansy as well. The plant has special meaning to my parents." Talfryn lifted the wilted yellow flower from the rocks. "Judging from the condition of the flower, she has been here within the last couple of days." He twisted to face the child. "Can you step back with Vendel so I can perform magic?"

"Mine now?" Signy held up the birch rune.

"For now," Talfryn reluctantly agreed, then waited for Vendel to take her hand and guide her away. Cupping the half-dead flower, he closed his eyes and thought of his secret name. Enchantments, restrained by the heavy cloak of his glamour, sparked within him. He willed the magic to rise upwards inside him. Darting like dragonflies, his fairy powers grew 'til he could contain them no longer. His eyes flew open. Staring into the dark depths of the well, he commanded, "Show me my mother, Disa."

Talfryn's breath hitched when he spied his mother's face shimmering in the water. Her wolf gray eyes appeared exhausted and her flaxen hair disheveled.

"She's alive!" Hope, joy, and relief streaked across his heart like a comet.

But as the view sharpened, hope turned to shock.

Chapter Forty-One
A New Firestarter

Talfryn reeled back on his heels.

"What's wrong?" Vendel released his hold on Signy.

"My mother's real… Older."

"Of course, she is!" Vendel thumped the blackthorn on the ground. "*Where* is she?"

"A farmhouse." Talfryn rose on shaky legs. "Sheep in a pen out front. I caught only a glimpse afore she disappeared inside."

"That farmhouse?" Vendel pointed. Far off to the right, shadowy figures called out to each other and laughed as they strode away from a blazing farm.

"Nay." Talfryn peered through the dark. "I see no sheep."

"Then leap ahead. Find the farm. Recover the potion."

"I can't leave you and Signy. The war band—"

"Is headed towards yer mother. There is nay time! Act now while ye have night's cover!"

Talfryn glanced at the burning abode. Already, the five men had disappeared into the darkness, their victorious boasts growing distant.

"Away with ye!" The dwarf shook the blackthorn. "I'll find ye."

"Stay safe. Do *not* abandon Signy."

"I vow she'll not be forsaken. Begone!"

Without further ado, Talfryn ran. Muscles bunched, he launched into the air. Smoke and ash stung his eyes. He blinked back tears, and after a fair distance, touched down, then leapt again.

The treeless dark heath offered little cover. Mid leap he spotted five shadowy figures barely illuminated by the low flickering torches they held aloft. Talfryn dropped quietly to the ground behind them. He held back, watching the dancing light glint off the Norsemen's Cold Iron ring-shirts and circular shields. But then his gaze fixed on the embers sparking from the tips of their torches.

A vengeful urge rose like a vortex inside him. His muscles corded, and heat flushed his skin. Magic flared down his sinewy arms, tingling his flesh and twitching his fingers. Jamming his hand into his pocket, he scooped up the last of the pixie dust. *This better be enough to keep them felled for a long time.*

Judging the distance separating him from the firestarters, Talfryn realized he would need to close the expanse. He could ill afford to have the magic dust fall short. With a quick intake of air, he sprinted forward, then jumped. *Surprise them. Leap close enough to slam them with as much dust as possible without being poisoned by their Cold Iron.*

Talfryn had halved the distance to the marauders when pixie dust began to rattle and crackle inside his fist as he raised his throwing arm. *Wait...* The acrid smell of smoke and Cold Iron burned a raw path down the back of his throat as he neared. *Closer...* The stench of Cold Iron grew even stronger. *Now!* Descending, Talfryn hurled the enchanted dust. Freed, it whizzed down in a spate of brightly colored specks.

He hadn't counted on the Norsemen's keen senses. Two whirled at the buzzing. Seeing him drop from above, they raised their painted shields. Pixie dust bounced off the wood and Cold Iron rimmed discs as the other firestarters, not as quick and thus unprotected, were struck. They thudded to the ground as though dead. The last of the falling magic specks smothered their torches.

Then the two remaining Norsemen flung their firebrands and drew their weapons. One unsheathed a sword. The other heaved a bloodied broad axe from his waist belt. With a roar, they lunged as Talfryn hit the ground. Without faltering, he leapt high and wide to evade their sharp-edged weapons.

On instinct, he launched a whirlwind—not nearly as powerful as an equistragon's deadly twister, but forceful enough to flip the sword wielder off his feet. The man crashed to the ground with a groan. Whirling air tore the sword from his grasp and sent it flying into the darkness. As the wind dissipated, Talfryn dropped to his knee on a wet patch, then lobbed a Moonstruck spell. When the white spinning ball hit the foe's chest and exploded, the man yelped and clutched his eyes. *Good. He's temporarily blinded. He'll be too dazed to move.*

Seeing the last of his band brothers defeated, the axe bearer uttered a ferocious growl and swung his weapon high. A fiery orange glow reflected off the axe blade. With horror, Talfryn noticed the sudden heat and realized his whirlwind had fanned the remaining firebrands. *I've set fire to the moor!*

Time suspended. All sound vanished.

The invader stormed towards him, but appeared to move slowly, in a time outside of time. As Talfryn leapt out of the way at the last second, time sped up. The attacker's yell, along with the fire's crackling hunger, assaulted Talfryn's ears. His opponent ran well past the spot where Talfryn had stood, apparently moving with such force, he could not halt. With a curse, the Viking pitched forward a few more feet, then abruptly sank in the mire to his knee while his axe slashed into the bog. Muck swallowed its blade and half the handle. Talfryn hit him with a Moonstruck spell, and the man stilled.

Flames jumped from one clump of bog grass to another, hopping over hidden marshes. Talfryn gaped at the blaze gathering force around him. The growing inferno closed in on the spell-struck Norsemen. Tunneling his fingers through his hair, Talfryn hesitated. *I should save them. But if I lift the spells, these torchbearers will continue their reign of terror. They will try to kill me. Yet if I do nothing, these men will die.*

Part of him thought they deserved to perish for the destruction they had wrought, and for what they or others in their war party had done to Signy and her family. *They are murderers! No different from Merrybelle's murderer.*

His ponderings were interrupted when thick smoke blew across his vision, and he became copiously aware the wild wind had whipped the flames 'til the conflagration roared like Beltane bonfires. Talfryn realized the capricious currents could carry the flames in any direction. *If the wind shifts, the blaze will engulf me.* With a sick feeling, Talfryn thought of the innocent humans and magical folk his fury was endangering. *I have become what the oak feared—a Norse firestarter.*

Mayhap the banshee's wail was not for me; mayhap it was because of me.

Chapter Forty-Two
The Summoner

The Blackwood Stronghold

Lord Argonshire must know I am marked with the taint. Fenella whirled and pressed her back against the locked chamber door. Panic rising, thoughts darting, she struggled to halt the tremors that shuddered through her. The past crashed over her, a breaking wave. Plunged back to the day she had been stolen and first locked in the tower, she relived the oozing blisters from being branded, the fever blazing her body, and the delirium that followed. Fenella hugged her torso; a strangled sob escaped her afore she could muffle it.

"Thaness?" Puffball made a small scratching sound as she scurried across the stone floor. She patted Fenella's foot. "Don't cry. Derry will release us."

Fenella dragged her mind back to the present. "Nay." She pushed away from the door. "By now, Argonshire will have told her I am asleep, and she is nae needed."

"Then your betrothed. He—"

"Edward must be injured or otherwise detained. He would nae allow this." Fear for Edward's wellbeing sent Fenella fumbling for the edge of her mantle. She wrapped the scarlet wool around her scarred hand.

"Then use your time wisely. Gaze into the water. Scry."

Fenella shook her head. "I accidentally scryed earlier, and it landed me here. 'Tis dangerous, even blasphemous." Memories collided within her, thoughts of her past sufferings pressing in upon fears of her present danger. Her head ached from mounting rage, betrayal, and hurt. Tears welled, blurring her vision. But she knew if she let them flow again, they would never stop.

Remembering her first days in this chamber and in what manner Edward had saved her after Nurse's ministering had nearly killed her, she thought how, over the past year, Edward had spent all his free time with her in Hamish's garden. *Our garden.*

"Talfryn is in danger." Puffball was tugging on her hem. "Quickly, afore that wretched nurse returns. Help him!"

The pixie's persistent plea broke through her reverie. So that after a moment, Fenella was able to still her mind enough to picture Talfryn as last she saw him reflected in the goblet's liquid. The skin on her shoulders began to crawl.

"You sense the danger, too!" Puffball led Fenella by the hem to the spilled water.

"Listen by the door for Nurse's return."

With a brisk salute, Puffball took up her post.

Fenella closed her eyes and gathered her wits. *Let nay harm come of this,* she silently prayed. Reluctantly, she opened her eyes, knelt, and peered at the puddle. "Heed me water. Show me Talfryn."

A blur of color swirled, then flared into focus. "Flames! Talfryn is surrounded by fire!" Fenella clutched her Thor's hammer necklace. "And I see bodies strewn on the ground around him."

"Bodies?" Puffball dashed to her side. "Is Vendel with him? Where is the human child?"

Fenella waved Puffball away, and the pixie scurried back to the door. "I see neither Vendel nor the wee lass." Her voice cracked. She stared harder. "The others appear dead. All men. Only Talfryn remains standing."

Puffball scurried back half way. "Did Talfryn murder them? He couldn't have. Could he? Or did the fire kill them?"

"No one appears burned. Though the blaze is closing in on them all." She twisted her mantle. "They seem to be on an open moor."

"Then Talfryn will use a Hydor spell to bring rain." A cautious sliver of hope crept into Puffball's tone.

"But the flames are so high. Will the spell be enough?"

Puffball slipped closer still. "I hope so."

The vision sharpened. Fenella smelled smoke. Her pulse jumped when the fire crackled and roared. Heat from the flames beaded her face with sweat. She sensed Talfryn's despair, his rapid heartbeat. *The men on the ground are nae dead. They are spell-struck.* The flames shifted, and she noticed a man in a ring-shirt, motionless, half upright, his torso curiously angled. She flashed on the time her cousin had stepped into a bog, sunk up to his knee, and became mired.

"They are on a marsh." Compelled by a force welling within her, Fenella extended her arms over the spill with her palms facing the water. A pulse awakened in the center of each hand. Its needle-like twinges grew strong and steady. Her eyes closed as an unknown spell came to her and escaped her lips.

"Water hidden. Water deep.
Waken now from your sleep."

Her hands grew icy as she became one with the dark water.

"Arise, arise. Flood the fire.
Heed me now, as I require.
Spare the Halfling and his foe.
Do nae let this wildfire grow."

Fenella's fingers trembled. The spill on the floor swelled. Water beneath the moor gurgled and bubbled. Peat splintered. A noisome smell of decay invaded Fenella's bedchamber. Magic pricked her fingers. As she raised her arms level with her shoulders, bog water rose through the cracks to engulf the moor. By will and by the twitch of her fingertips, Fenella tried to direct the flow around the spell-struck men.

"Control your magic." Puffball scampered to her side; her drawn-out tone urging Fenella as if she were somehow helping.

Feeling water soaking into her léine, Fenella realized murky bog water had reached beneath the chins of the men prone on the ground.

Puffball scampered up Fenella's arm to avoid the expanding puddle. Settled beneath her hair, the pixie's quick breath prickled her neck. "Don't drown them!"

With alarm, Fenella fisted her hands to halt the water's surge, and the spill quickly shrank.

"I dinna douse all the flames." Fenella set to completing the job. But her arms bobbed when she caught sight of Talfryn's bewildered expression afore he raised his arms and shouted. She heard *"Hydor!"* Then after a moment, rain fell. *That will help.*

Absorbed in her task, Fenella did nae hear Puffball's warning nor the key in the lock. She did nae notice the cool draft when Nurse opened the chamber door nor the earthy scent of the sleeping draft. Only the clatter of a dropped tray and the heavy thud of a body collapsing on the floor behind her pulled her from her vision.

Fenella stared at her hands, then lowered her arms to her sides. Her stomach whirled as she shifted from the fiery moor in a distant time to her chamber in the present.

Behind her, Puffball cleared her throat, a sound like autumn leaves rustling in the wind.

As Fenella rose, she turned towards the commotion that had broken her concentration. Clumps of tall grass poked out from beneath Nurse's prone form. "You cast Stray Sod?"

"She saw you." Puffball's defensive tone rose. "And me! I had to stupefy her."

Reaching over the fallen woman, Fenella pushed the heavy door closed.

"I wasn't sure the seeds would sprout on the stone floor, so I sowed a lot," Puffball confessed. "Did I kill her?"

"I hope nae." Fenella crouched beside Nurse and placed her finger beneath the woman's nose. Nurse's breath fluttered against her skin. Fenella exhaled as well. "She's alive."

"We need to flee afore she wakes up."

"Fleeing will nae solve our problem. Even if she is stupefied and canna remember what she saw, his lordship will be suspicious. The grass was nae here afore. He'll think I performed magic."

"Then I will magic away the grass, hope the woman was on it long enough to forget what she saw, and make her fall look like an accident. Then we flee."

Glancing at the floor, Fenella noticed much of the sleeping potion remained in the flask. Strengthened by her water working, she opened her mind to magical thoughts. A plan soon flowed to her.

With a twinge, she wished it could include Edward.

Chapter Forty-Three
Moral Dilemmas

How could I have caused this blaze? Fire roared steadily closer. Crackling flames and searing, breath-stealing smoke transported Talfryn back to Merrybelle's death, to Lily's scorched wings and unhealable burns. *If any magical creatures die because of me, Lily will think me a killer. But after Merrybelle, would Lily save these firestarters? Surely, she wouldn't want me to become like them. If she knew I was responsible for this firestorm, she would want me to do whatever I could to prevent more death, wouldn't she?*

As if in answer, the ground beneath him trembled. All at once, unseen pockets of bog water announced themselves with gurgles and blurps so loud, Talfryn heard them over the conflagration. Water once below the surface, soaked into the soles of his soft leather boots and surged around the fallen Vikings. Hisses punctuated the air as the cold, smelly water doused much of the fire at its source.

Magic! Talfryn peered through the sun-bright flames, half expecting to see a water mage or a Mistress of the Bog. The peat beneath him cracked open. Arms swinging, he jumped off the widening gash and splashed down next to one of the moonstruck war band. Closed in by towering flames, he raised his arms to the stars. *"Hydor!"*

One heart stopping moment passed. Nothing happened. *If the rising water doesn't halt soon, they'll die by drowning instead of by fire.* Talfryn kept his arms outstretched.

A light shower, summoned by his Hydor spell, fell in a balmy sheet over the bog and quenched the remaining flames. Upon the rain's arrival, the bog water receded as abruptly and mysteriously as it had risen. Raindrops dripped off the shivering, spell-struck men and pattered onto the drenched ground.

Frowning down on the helpless lot, Talfryn felt his father's thunderstone claw weigh heavily against his chest. "May the Fates decide your future," he told the war band. Then, fearing for his mother, he set off with a running leap.

****

At the first sign of the dragon warships, Disa and Torolf had killed the fire in the hearth and extinguished the candles, plunging the stone farmhouse into semi-darkness. Twilight ghosted in from the smoke hole cut into the sod roof. Beyond the faint shaft of light, oppressive gloom filled the barren room. The low door had been hung to open outwards so it could not crash inwards during violent weather. Disa hoped it would also leave intruders exposed and vulnerable if they overtook it.

Pacing to keep warm, she struggled to silence her nine-year-old self who screamed, *Flee! Torchbearers are on the hunt! They are nay different than the slave raiders who burned down our home and killed—*

Abruptly, Disa halted. *I am a shield-maiden now, not a helpless child. Nay spirit bear huffs at my back, waiting for me to die.*

She looked over at Torolf. Two days after he had found her on the moor, his wife's runes had foretold the warships' arrival. *Thank the gods and goddesses Torolf hastened Idony and their two children off the island,* Disa thought. *But why did he so foolishly return?*

"You should have stayed with your family," Disa said in a voice just above a whisper. "I can fend for myself."

"Aye. You can." Torolf crossed the dark open room and placed a small sack of provisions near the door. "But two blades are better than one. I cannot help you penetrate the magical wards the way Idony might, but I can protect your back."

"I am grateful for your aid." Her eyes had grown accustomed to the dimness. What she could not see of Torolf, she filled in from memory. His long blond hair matched her own in color and length. Idony had woven his woolen cap. The sewn rabbit pelts he wore over his tunic hung almost to his knees, half covering his woolen leggings and cross garters. Silver-and-gold wire, fashioned into intricate designs, decorated the wide leather belt he wore over the furs. His eyes, she realized, had been in shadow since his return.

Guilt entangled her. Torolf and Idony had showed her more kindness in the past fortnight than she had seen in over eighteen years. Avoiding the magical moonstone inlay, she brushed her fingers across the cold hilt of her sword. *For the first time since Beacon was at my side, I trusted others enough to sleep through the night.* A rush of elation shivered through her as she remembered the dreams that had come to her. *My first dreams, my first deep sleep since Mab banished me. Dreams!* She frowned, struck by guilt. *Torolf and Idony know naught of the Aeternitas Potion or that I was—am 'til Grambolyn frees me—a slave. He knows naught of who I truly am, and yet he returned to aid me.*

I am so close to finding my way back to Grambolyn; so close to the dwarf fulfilling his promise to find Bree. She smiled, imagining embracing her sister again. Bree's eyes would twinkle like Muirgen's, the fairy cat. Disa could almost feel freedom glide like the downy fairy hair of her son across her skin. *Oh, to be free of this crystal vial I have safeguarded for so many years.*

Her core chilled. Regret rose unbidden, sudden, and painful, like a dagger to the chest. *Beacon.* She could still see the misery in his eyes when she had exacted his promise to not search for her. Their baby, so sweet and fay like his father…must be fully grown by now. The pain in her chest sharpened. Blinking back tears, she walled off the despair threatening to breach her defenses.

I have survived this long on my own. I can survive a few more hours.

Disa envisioned the standing stones she had traveled so far north to reach. So many stones! She remembered the odd disturbance that had rippled twice through the farmhouse today.

"Do you think the gods wish me ill?"

"Because of earlier?" Torolf asked.

"Aye, when you arrived."

Like a brother, Torolf clasped her shoulders. "If the gods wished you ill, they would not have brought you to our door."

Disa blinked, reassured. Surely Fate, through the blessings of the goddesses Hlín and Frigg, had led her to Torolf and Idony. She had won the couple's trust and hospitality by telling them half-truths that served her purpose; that she had been beguiled by a fairy and later stolen away to his realm where she gave birth to a son and was then banished. *"I can never return to Fairy or see my child,"* she had told them, true tears trickling down her cheeks. *"I have been searching for a way to return to my own time, to those who raised me. Along my journey, I have heard the standing stones may have the power to magic me home."*

Torolf released her shoulders, returning Disa to the present.

"The ripples were some sort of magic," she said.

"Aye. Protective. Druidic or earlier."

"I have heard the Traveler's Moon arrives once every eighteen and a half years. But none of the tales mentioned a spell launch."

"Idony's runes foretold a great terror and conflict on the island would trip ancient magic. Since you intend nay harm, you will be safe."

"But can the magic discern between the innocent and those with ill intent?" Disa worried out loud.

"It passed through us, so aye. Besides, I sense the magic has run its course." Torolf strode to the wall facing the sheep pen and path to the front door, then removed a wad of straw from a gap between the stones. Pressing his eye to the hole, he said, "'Tis gloaming. A ground mist is creeping in."

"See any invaders?" Disa whispered.

"Nay. The moon is about to rise."

"We should leave soon." Disa rubbed her arms to ward off nerves and the chill. "Let us free the sheep, then depart. I would rather battle ancient magic than wait here for torchbearers."

"I doubt we will be discovered. The berm hides the hut, and the roof is barely discernible from the meadow."

"Aye, but war bands might hear the sheep, should something disturb the beasts. How much longer 'til moonset?" Disa asked.

"Difficult to predict during a Traveller's Moon. Two hours. Mayhap five." He turned his back to her and stared out once more.

Disa quickly raised her cloak and felt for the crystal vial. The cloth seam felt warm beneath her fingers, as though the hidden potion radiated heat. *How odd. Heat, not cold?* Dropping the cloak's hem, she worried her hand beneath her padded leather over-tunic. Finding the hidden pocket, she felt for the small crystal sphere the dwarf mage had given her.

"Its power will only work once," the mage had told her. *"When the moon shines in exactly the right place and your heart is sure."*

After all these years, Disa was finally certain she would catch the moon at the right time, in the right place. But having offered up her goddess rune at the sacred willow well, her stomach knotted. *I have never faced danger without Hlíne's charm.*

Have faith. After almost nineteen winters, I will finally return home, be freed, and find Bree. Then I will search for Beacon and Talfryn.

Outside, the sheep bleated. Disa stilled, her muscles and nerves tight, her senses heightened. Torolf peered through the gap. He turned, faced her, and raised his forefinger.

One intruder.

They reached for their shields and helmets.

Torolf slid the axe from his belt, then angled his head towards the door.

Disa eased her sword from its sheath. Clutching the hilt with one hand and her shield with the other, she crept towards the wooden door. Torolf edged along the wall, his axe ready.

Will the intruder try to kick in the door, or worse, set it afire?

Instead, an unfamiliar voice called her name. "Disa?"

Disa glanced at Torolf. He furrowed his brow and lifted his shoulders in wordless puzzlement. She shook her head. They both crept closer to the door, their weapons drawn.

"Disa, wife of Beacon, Head of the Noble House of Storms!" The voice, youthful and masculine, called out again. "Are you within?"

Leading with her sword, her heart thundering, Disa slowly opened the door.

Chapter Forty-Four
Reunion

Talfryn heard the creak of the low wooden door open as a sword tip slid through the crack.

"I seek the Norsewoman, Disa." *Am I too late?* His heart sped up like fiddle and hornpipe music during a full moon dance.

He had almost missed the low, oval farmhouse, cleverly hidden behind a berm and topped with a sod roof. Had the sun's last rays not glanced across the sheep in their hidden pen, he would have leapt past it. His sudden appearance had set the sheep crowding and bleating. Certain he had landed in the right garden next to the pen where he had seen his mother, he now waited, hoping it was her behind that door.

Shivering from nerves, he drew his cloak tightly around his body. The breeze iced, carrying the musty, metallic stench of Cold Iron. His chest tightened, and his skin beneath the serpent bracelet burned.

When the door finally widened, the double-edged blade emerged from the shadows, and the woman holding the sword pointed its Cold Iron tip towards Talfryn's heart. Beneath her Cold Iron helmet, her gaze darted to the path behind him. Apparently seeing no one, she trained her steely gaze on him.

Over the Cold Iron band encircling her shield, Talfryn met the woman's fierce eyes—the same wolf gray as his own. All the words he'd dreamed of telling her clumped in his throat. He swallowed them down like moldy leaves on a damp forest floor.

Too unnerved to speak, he stepped back. His vision blurred. Was it the Cold Iron or the sudden, unexpected tears that welled his eyes? Mayhap both.

Behind Disa, an imposing figure ducked through the shadowy doorway, axe in hand. The man spoke in Old Norse, his words not registering, only his tone, suspicious and deadly, cut through Talfryn's shock.

"I am Talfryn." He raised his hands to prove he carried nay weapons, then directed his next words at his mother. "Your son."

The shield-maiden's sword wobbled. Her shield dipped. She blinked several times. Talfryn, watching her lips and chin quiver, feared she would cry. Then Disa appeared to gather herself. Hope seemed to battle suspicion in her eyes. Could she hear the erratic beat of his heart?

He touched the circular brooch pinned near his shoulder. His fairy cloak fell open, revealing his mother's Norse armband and the serpent bracelet coiled around his wrist. The ruby glow from the serpent's eyes glinted off Disa's sword tip like a spark.

Her eyes widened. Disa's gaze darted between the brooch and armband, both gifts he knew she had bestowed upon him minutes afore she was banished. Suspicion seemed to fade from her eyes as her expression crumbled. Talfryn noticed the shaking that had begun in her arms traveled like an earth tremor through her body.

Disa swayed and reached for the man at her side. The stranger clasped her elbow as she lowered her sword. Her strong grip now appeared boneless. The gleaming sword clattered to the ground.

Talfryn stepped towards his mother, but the Cold Iron forced him back.

The Norseman glanced at Disa with obvious concern, his focus shifting between mother and son.

She tilted her shield against the wall behind her. "Torolf, lower your weapon," Disa instructed him in Old Norse.

"Are you certain?" Torolf asked.

"Aye."

Torolf shoved his axe handle into his belt and stepped back.

As though to stem sobs threatening to escape, Disa pressed her fingertips to her lips. She sniffed, then closed the chasm between them and reached for Talfryn's hand. An ache bloomed in his chest when he noticed how much older she appeared than his father, who had lived for centuries without aging. Instead of flawless fairy skin, her face bespoke harsh winters and years of worry.

Smoothing her thumb over the ring he wore, Disa asked in accented English, "Whose ring is this?"

"My father's." Dizzy from the Cold Iron, Talfryn fumbled as he slipped the ring from his finger and threaded it over her thumb. His voice cracked when he noticed the ring of gold, blue tourmaline, and two smaller stones on her finger. "He bade me give it to you so you'd know he had sent me."

Disa's lips trembled. "Talfryn? 'Tis truly you?"

"Aye, Mother."

"I feared Mab had—" She shook her head as though to clear terrifying thoughts. With a sob, Disa embraced him so tightly he could barely breathe. She smelled faintly of the sea, smoke, fish, and Cold Iron. When she ended the hug, she cradled his face in her hands. "You appeared so fay when you were born. I never expected—"

Talfryn broke from her grasp and staggered back. He felt the blood and color drain from his face. His legs threatened to buckle.

"Son?"

"Cold Iron," he rasped.

"Forgive me! You appear so human; I did not realize." Removing her helmet, Disa retreated to the doorway. Switching to Old Norse, she spoke to the hulking farmer in urgent tones. Torolf cast Talfryn an astonished glance. Disa removed her belt and a hidden dagger. Then, at her behest, Torolf carried the Cold Iron into the farmhouse.

A modicum of strength returned to Talfryn's knees, but his head remained dizzyingly light.

"Better?" Disa asked as she approached him.

"A little." *Concentrate.* "Mab believes you have the Aeternitas Potion," he said in a wavering rush. His mother grasped his arm in a surprisingly strong grip. "Mab holds Father spellbound and tethered to Cold Iron. She is killing him slowly. Painfully. He will die if I do not return soon with the potion."

Disa covered her gasp with her hand, then whispered through her fingers, "We must save him! I told Torolf about you and your father, but he knows nothing of the potion."

Afore she could say more, Torolf returned, appearing uncomfortable without his helmet and shield. His axe remained belted to his waist. Talfryn wondered how many other weapons were concealed beneath his clothes.

Clearing her throat, Disa gestured towards the farmer. "Talfryn, this is Torolf." Her voice was husky with emotion. "Mab banished me nearly two centuries into the future from my own time. Recently, Fate with the help of the goddess Frigg led me to this island, to Torolf and his wife, Idony. Like me, they practice the Old Ways." She switched to Old Norse. "Torolf, the gods and goddesses have returned my son to me. Behold, Talfryn."

Talfryn stepped forward and the two clasped forearms. Their gazes locked, and the farmer flinched as though he recognized him. Talfryn trenched his brows, certain he had never met the man. Torolf cast a curious glance at the glowing serpent bracelet, released his grip, then stepped back.

"'Tis a sign, já? Your son appears as you are about to return to your own time," Torolf said.

"Aye." Disa averted her gaze.

"You have found a way back?" Talfryn asked.

"I believe so. We must hasten, or I will miss my chance. Come with me."

"Nay. I must return to Fairy with…the object Mab sent me to retrieve."

"First, we bring it to the dwarf mage. Gods willing, she is still alive. Surely she can conjure a look-alike potion and vial, which we can bring to Fairy to save Beacon."

"There is no time," Talfryn said, his voice rising. "Mab will not lose her immortality if she kills a human. If she thinks you stole from her, consumed some Aeternitas potion, and acquired immortality, she will murder you, then focus her wrath on Father. He is probably half-dead by now." Talfryn shuddered. "If we delay, or deliver less than all of the potion, then Father's and our fates are sealed."

Disa winced.

Torolf spoke up. "We must leave lest you miss the moonset. I will retrieve our weapons." After a pointed glance at Talfryn, Torolf stalked back inside the hut.

"Ye found her!" came a call from the curve in the path. Vendel huffed towards them, his eyes watering as he jabbed the blackthorn staff, now dusted white with ash from the burnt areas he had traveled through. He carried Signy on his back. "Disa, do ye remember me? 'Tis Vendel Fairheart."

"Hilde's mate?" Disa strode towards Vendel and clasped his forearm. "I feared I would never see another dwarf! You are still alive, so Mab must not know—"

"Mab suspects nothing. Centuries have passed with no overt action. She thinks we have forgotten about the Last Fairy King and Gudbrand."

"Ha! As if dwarfs would forget Gudbrand's murder. And Hilde?"

"Departed Fairy the morning Talfryn and I began our quest."

"Wise."

Watching her—them—Talfryn realized how little he knew about the dwarfs working the crystal mine. He shifted from one foot to the other. Like the enslaved human children, he had taken their presence for granted. *I thought little of what their lives must be like.*

"Signy?" Disa asked, taking notice of the human girl.

Talfryn whispered, "Father murdered. No sign of Signy's mother."

"Nay!" Disa pressed her palm to her heart. "I must inform Torolf."

"The bad news can wait." Vendel grabbed Disa's wrist, his tone harsh. "Do ye have the potion?"

"Vendel!" Hostile magic sparked from Talfryn's fingers. "Unhand her."

Signy kicked and demanded to be let down. She slipped from the dwarf's back to the ground. Talfryn subdued his crackling magic and coaxed the child to his side.

"Aye," Disa said, in a voice as deadly as Cold Iron. "I still possess the potion." She wrenched her wrist from Vendel's grasp.

"Signy?" Torolf exclaimed after ducking through the low door and absorbing the scene afore him. The farmer dropped Disa's weapons and a sack. He once again wore his helmet and carried a shield. His hand hovered near his axe handle as he scanned the path. "Where are Astrid and Dag?"

Disa strode to the doorway and turned her back so only Torolf could hear the fate of Signy's parents.

"Dear gods!" Torolf twisted away from her words.

"Signy's father was his friend," Disa explained, facing them again. "Their daughters..."

The little girl ran to Torolf and hugged his leg. The farmer swiped the tears from his eyes, tilted his shield against the hut, then lifted Signy and settled her on his hip with his arm encircling her.

The child stared into Torolf's eyes, then leaned away as though she wanted to be lowered to the ground.

Instead, Torolf wiggled his fingers afore the girl's eyes and recited a calming rhyme under his breath. Signy's small shoulders slumped, and she leaned against Torolf's chest.

Disa hurriedly reclaimed her weapons. As she fastened her sword belt, she took charge. "A portal will open at the standing stones at moonset. 'Tis not far, but once there, we may have to battle defensive magic." Switching to Old Norse, "Torolf, I thank you for your protection and aid. But I will accept nay more. You must leave us to our fates. Save yourself and the child."

"Are you certain?" Torolf asked, eyeing Vendel.

"Aye." She strode forward and clasped his forearm. "Take the provisions. Let your neighbor's wounds be your warning."

"May the gods and goddesses be with you." Torolf nodded to

Talfryn and Vendel, hefted Signy higher on his hip, then retrieved the sack and his shield afore setting off. He turned back once and caught Talfryn's eye. Torolf's form flickered. For a heartbeat, a young druid with unkempt brown hair, pale skin, narrow shoulders, and a pointy, goat-like beard appeared.

Merlin? Talfryn blinked. Torolf, again himself, strode off.

My eyes must be deceiving me. Still… Talfryn frowned. "Are you certain Signy is safe with him?"

Disa's brow furrowed. "Quite certain. You saw how she ran to him. Torolf is a good man. He and Idony will love and protect Signy as though she were their own."

As though she had heard her name, Signy glanced back. Solemn-eyed, she stared at Talfryn for a moment. She straightened and waved once, then twisted and faced forward towards her fate.

Frowning, Disa opened the gate to the sheep pen. "Away with you!" she said in a harsh whisper.

Talfryn tore his gaze from Signy and glanced at Vendel. The dwarf watched Disa's every move. Unease pitted in Talfryn's stomach at the stony set to Vendel's jaw and the flinty shine in his eyes.

Chapter Forty-Five
Past or Future

As Disa led the others, she cast furtive glances at her son. *How did such a fay babe grow up to be so muscular and appear so human? I thought he would resemble his father, slight of build, high cheekbones, pointed ears.*

Beacon. Following the orange Traveler's Moon as it skimmed distant hills on its journey towards the Callanish Stones, she could barely breathe imagining Beacon near similar stones and tethered to Cold Iron. Like bellows to a forge, her hatred for Queen Mab flared. *Afore, I only wanted to stop her. But now…* She glanced down at the sword belted to her waist, then pressed on harder, coming upon their destination just afore moonset.

Either by magic or by wind off the water, nay smoke obscured her view. Disa breathed in scents she remembered from the dwarf mage's magic long ago—dragon's blood and meadow rue. Studying the central circle of stones, her skin crawled as though a dozen beetles had crept from the forest and now skittered across her flesh.

The stones are alive! Disa leaned towards the upright figures. *I can almost see the cloaked souls they once were. An ancient race… mayhap ancestors of mages or druids. Gathered in ritual, then turned to stone by a misguided or vengeful spell? 'Tis of nay matter now. They became trapped within the stones.* She rubbed her chest where the Stone People's hostile life force pushed against her. From within the hidden pocket in her padded tunic, she felt the mage's crystal sphere tremble. Disa slipped her hand inside the pocket and stilled the stone.

What would the mage, or for that matter, Ragna, advise me?

Her thoughts jumped to her long-ago meeting in the seer's cave. Ragna, or rather the druid speaking through Ragna, had said Mab's greatest fear was the truth.

Disa halted. "Why now? Why does Mab urgently need the Aeternitas Potion?"

"A fairy named Merrybelle died in a fire that reeked of dark magic." Talfryn glanced at Vendel. "We suspect—"

"Mab caused her death." Disa barely contained her excitement. "How long ago was Merrybelle murdered?"

Talfryn winced.

"In Fairy time?" Not waiting for an answer, Vendel pulled out a tiny, gleaming object fashioned to resemble a miner's tool. When he twisted the pewter handle, the device whirred, and a crystal disc slid out. The dwarf stared down at the flat, glowing moonstone.

"Mab sent us to search for you right after Merrybelle died. The full moon was due in five days. So, four days ago."

"Traveling through time may have harmed the Luminositor." Talfryn frowned. "Or not."

"Aye, but what if four days *have* passed?" Disa clasped his arm. "'Twould be too late for Mab to use the potion."

"The killer will have started to age." Vendel's eyes gleamed.

Disa withdrew her hand and stared once more at the standing

stones. For the first time since childhood, she was uncertain of her next step. "I wish we knew if Beacon still lives."

"We must believe he does." Talfryn pulled a talon-shaped talisman from beneath the neckline of his tunic and rubbed the iridescent crystal.

Fear scrabbled down Disa's back. "Beacon gave you his thunderstone?"

Talfryn's fingers stilled. "For safekeeping."

"But he would not give it to you unless…" *He expected to die.*

Mayhap sensing her alarm, Talfryn tucked away the symbol bequeathed the head of the Noble House of Storms.

"Ye made a vow to Grambolyn. Yer choice is set." Vendel spoke in earnest, his tone harsh and insistent. "Show us to the standing stones."

"Did traveling through the smoke blind you?" Talfryn asked. The stones are right afore you."

Vendel's forehead crinkled. He leaned forward and peered.

Worry gnawed Disa's bones. *His eyes nay longer weep from his smoky journey, and the air is clear here. Yet, he cannot see the stones.* She thought back to Torolf's words. *"Idony's runes foretold a great terror and conflict on the island would trip ancient magic. Since you intend nay harm, you will be safe."*

As she pondered what harm Vendel might intend, she noticed a shimmering horizontal band take shape in the air surrounding the stone circle. *Another spell-halo?*

"My mother's choice is not set. She has her husband to consider." Talfryn folded his arms over his chest and switched his attention to her. "How old were you when Grambolyn bought you?"

Disa flinched as the question tore her from her thoughts. "Nine winters." Her shoulders rolled forward when she shrank into the dreaded memory. *The slave market. Men staring, judging, shouting. The bite of the iron shackles. Thank the gods Bree had been spared the tether. "Not that one." The slaver had pointed to Bree. "She is too timid to flee."* Disa rubbed her palm, remembering Bree's small, sweaty hand in hers as they were forced with the other captives into an empty animal pen to await their fate.

After the pen, I never saw Bree again. Oddly, though, as she was pushed onto the block and her back was to the pen, she had seen a flash of light out of the corner of her eye. For a second, she had wondered if her *fylgja*, her soul's protector, had come for her.

"You have safeguarded the potion for more than twice the number of years you were enslaved. Centuries have passed since Grambolyn devised this quest. He is long dead."

Her son's words jolted Disa back to the present. "The quest was my idea."

Talfryn's tone softened. "Grambolyn tricked you into believing it was your choice."

Disa reeled back. "Nay."

"No one tricked her," Vendel insisted.

The spell-halo around the standing stones thickened.

"Mother, you were not born a slave. Like my friend, Fenella, you were stolen. You—"

"What?" Ragna's parting words resounded in Disa's mind. *Remember, you were not born a slave.*

Talfryn faced Vendel. "Connell would have given Fenella the Finding Horn. It will show us Father!"

"Remember how long we had to wait for Fenella to leave her castle? 'Twould be faster to magic to Grambolyn's time," Vendel argued.

"Is the dwarf village free of Cold Iron?" Talfryn challenged.

"Nay," Vendel snapped his reply.

Cold Iron! Disa wavered. *Must I choose between endangering my son, or possibly creating further delay that may cause my husband's death, all on the slim possibility I will find my sister?*

Vendel rubbed his chest. "Do either of ye feel—"

Like a horse flicking away a fly, Talfryn shook his head. "Mother, if you return to Grambolyn, you will return a slave. You will be safe with Fenella. She is from the future. Her clan will protect you."

A young woman? From the future. "What color is Fenella's hair?"

Talfryn furrowed his brow. "Red. Why?"

The defensive magic gathering near Vendel thrummed her ears. Disa had nay idea what a Finding Horn was, but it sounded magical and useful. *Might Fenella be the ally Ragna had promised?* She grasped Talfryn's hand. "Take me to her!"

"Ye swore to help the dwarfs avenge the Last Fairy King and Gudbrand. If ye won't, I—" Vendel reached for her cloak. Magic rippling off the standing stones barreled into him and knocked him sideways.

Talfryn seized Vendel's wrist. "Mother, complete the circle!"

Disa grabbed the dwarf's free hand as her son's next words were lost beneath the drumbeats emanating from the spell-halo. She narrowed her eyes at Vendel. *I have been alone so long, I nearly forgot Grambolyn's most important lesson: Dwarfs never forget an injustice and are single-minded in their quest for revenge.*

Generation, after generation, after generation…

Chapter Forty-Six
Argonshire

The Blackwood Stronghold

Fenella had failed to remember that this close to the solstice, the sky would still be light despite the advanced hour. Thankfully, the grayish-brown hood of Nurse's dun cloak concealed her bright hair and shielded much of her face as she crossed the upper courtyard, holding the garment closed to cover the fairy cloak tied to her waist. She dared nae run. Instead, imitating Nurse's purposeful strides, Fenella kept her back to the guards on the ramparts and hastened towards the gate.

Hoping to spot Uncle Connell, or Sorley in his wolfhound form, she scanned the courtyard. *Empty*. Puffball, hidden within an herb pouch tied beneath Fenella's cloak, rocked against her.

Her shallow breaths, her footfalls, and the iron keys jangling in her pocket sounded overly loud to her ears. Any moment, she expected to hear a guard raise the alarm.

Sudden flashes of color in her side vision startled her.

Thud. Thud. Thud.

She halted and swung towards the disturbance. *Talfryn! Vendel! And the blonde!* Fenella's pulse faltered.

Talfryn slumped to the ground.

"Thaness?" Puffball whispered from her hiding place.

"'Tis Talfryn."

The serpent bracelet's scarlet glow cast an eerie light through the Halfling's cloak.

Nay! Fenella hurried to Talfryn and knelt beside him. "Cold Iron?"

Talfryn blinked.

"You still appear Norse."

"Glimmer's glamour…holds tight." He reached for her with a feathery grip. "Tell me you have Puffball."

The pixie crawled out of the pouch, then flipped back the dowdy cloak. "You huffed off without me!"

"Puffball, thank Mab, you are alive!" Talfryn's hand trembled.

And you escaped the fire. Fenella softened her expression when she gazed into the blonde's gray eyes, the same gray as Talfryn's. "You must be Disa. I am Fenella, Thaness of Thorburn."

"Our meeting was foretold." Disa cast a worried glance at the stronghold.

"My son needs a safe place to recover."

Talfryn struggled into a sitting position.

Puffball's leafy-green ears twitched. "The air beneath those trees is free of Cold Iron. Avoid the byre and stronghold. We must make haste afore Fenella's absence is discovered."

"Were you fleeing?" Talfryn raised his arms so Fenella and Disa could pull him to his feet. "And why are you dressed thus?"

"'Tis my disguise," Fenella replied as they hurried towards the few trees near the wall. "We were going to search for the Romani."

"But your wedding—"

"There are steps I must take first. Why are you here?"

"To find Puffball, and I had hoped to leave my mother with you while I return the Aeternitas Potion."

"We need a new plan," Disa said.

"My apologies," Fenella said.

Talfryn peered around a tree. "We are walled in."

"I stole a key ring," Fenella answered. "One of the keys might unlock the gate."

Puffball removed her cream skullcap. "Hand them to me. I will be harder to spot from the ramparts."

"But the keys are made of Cold Iron," Fenella protested.

"Pixie. Better resistance, remember?" Puffball snatched the heavy ring, then dragged the unwieldy bundle as she bounded across the grassy courtyard. The keys clattered as the pixie scaled the gate. The first key did nae fit the lock. Fenella bit her lip. Puffball tried the second key, then hefted the third, and finally the fourth. She shook her head, then scrambled to the ground and dashed back to the group.

"No luck," Puffball whispered, huffing when she returned. She handed Fenella the keys. "Now what?"

"Do you have the Finding Horn?" Talfryn asked.

Fenella patted her hip. "Aye."

"Magic us to Grambolyn. We can use the Horn there," Vendel insisted.

Talfryn glared at the dwarf. "I'll not chance my mother and Fenella being enslaved."

"Hide us in Fairy," Disa suggested.

"And risk your death? Nay."

"The fairies have a pact with my clan," Fenella reminded him. "Your mother should be safe as long as she is under my protection."

"Mab won't honor the pact," Puffball said, scrambling up to Talfryn's shoulder. "Not under these circumstances."

"To the Romani then," Fenella said, pleased with her solution that would meet both their goals.

Shouts rose from the castle. Guards stared over the parapet.

Fenella's pulse jumped when she saw Edward, alone, dash down the outer timber stairs, then angle towards the byre. Concern and anger warred within her as she fumbled the keys.

"The Blackwood is getting closer, and he has Cold Iron," Vendel warned.

Puffball tugged on Talfryn's ear. "Make haste!" She dove into his cloak pocket.

Edward drew near. When their gazes collided, Fenella flashed on their meetings in the secret garden, then to the click of the key in the lock.

Then Talfryn shifted closer to her, seizing Edward's attention. The Scot halted. Fenella's heart tumbled as Edward's wide-eyed surprise shifted to a pained stare.

Talfryn clasped her hand, causing the keys to slip and jangle as she caught them in her free palm. She saw him reach for Disa. Uncertain, Fenella regarded Edward.

In seeming desperation, Edward stepped closer. Bruises swelled one corner of his mouth. "I returned to your chamber to free you!"

"But then who raised the alarm?" Fenella asked.

"A guard must have seen you flee. Stay, and I will protect you. Or we can leave now, together."

Fenella wavered. Nurse's cloak snapped in the rising wind.

Edward limped forward. "My heart will nae beat without you."

The heartache and longing in his voice stole Fenella's breath. She broke from the circle.

"Don't let him sway you!" Talfryn said, grabbing her arm.

"Unhand me!" Fenella insisted.

Edward shortened the distance between them, but afore Fenella could shake off Talfryn, Disa gripped her other arm. Vendel, his jaw set, completed the circle.

"Nay!" Fenella screamed. A current zinged through her, filling her with dread. "Edward!"

Blackwood lunged, but only grasped keys as the magic ripped her away.

Chapter Forty-Seven
Epiphanies

Shock shook Talfryn to the innermost crevices of his being. *Is that what human love looks like?* The Blackwood's fierce expressions of dismay, then determination replayed in Talfryn's mind. Absorbed, he ignored the fraction of his mind that noticed a strong force was blocking his magic and veering them off course.

Blackwood was trying to save Fenella from me. Me! The thought knocked against his skull as his feet slammed the ground. He landed off balance and tightened his hold on Fenella to steady himself. Bracing sea air filled his lungs. *This is not the Romani camp.*

"Where are we?" Puffball asked.

"I don't know. A headland. But where? My intention was the Romani camp." Disoriented, and dizzy from his mother's Cold Iron, Talfryn staggered back a step.

"*Someone* brought us here." Fenella broke free of Disa. With a chill, Talfryn saw Blackwood's raw emotions mirrored in Fenella's eyes. "Unhand me," she demanded.

Talfryn released her. He set his jaw and pushed aside a stab of guilt. Would Blackwood have declared his love for Fenella had he known she was—what had the Scot called magical ability? *Tainted? Nay! He wouldn't have.*

Fenella compressed her trembling lips, and for a terrible moment, he feared she might sob. His heart wrenched.

But when Fenella's expression hardened, he sensed magic gathering within her like a storm at sea. The breeze lifted her bright red hair, reminding him of Viviane, the Lady of the Lake, and Morgan le Fay. A prickle of unease danced across his shoulders.

"Fen—"

She raised her palm to him, and his jaw clamped shut.

Fenella strode three long steps, crushing buttercups and purple woundwort afore halting at a stone outcrop overgrown with heather. Standing near the tip of the promontory, she stared at the water, spread her arms, and unfurled her fists.

"Water magic…" Vendel warned, squinting into the distance. "The lass is churning the sea!"

"By the gods!" Disa gaped at the water, then pressed her fingers to her lips.

Talfryn's attention jumped from Fenella to the swirling sea, then back to Fenella.

High color blotched her freckled cheeks. Tears fled her eyes, leaving glistening trails. She swiped them away with the side of her hand, then dropped her arms to her sides. With a seething glance at him, she raised her chin, then angled away.

"It was you," he whispered.

She returned her gaze to the frothing waves, and the sea calmed.

"You raised the bog to squelch the flames. How did you know I needed aid?"

"Seriously? You have to ask?" Puffball rolled her eyes.

Fenella stood silent.

"We cannot delay while Beacon yet suffers." Disa's intense stare traveled from the blue-gray sea to Fenella. As if feeling the heat of the shield-maiden's attention, Fenella shifted and met Disa's gaze.

"Thaness, my husband's life is in your hands. Your magic surpasses even the dwarf mage's power. But I sense your abilities are newfound, já?"

Fenella nodded.

"Unschooled and uncontrolled magic endangers us all."

Fenella visibly tensed.

"Surely one of your stature has many skills that would aid our cause," Disa continued. "Fate led us to you for a reason. Pray, prove the gods and goddesses were wise to choose you."

Talfryn shifted from one leg to the other, avoiding Fenella's sideways glance.

"Son," Disa said from a safe distance. "You mentioned a Finding Horn."

"Aye." A sudden awareness, a tug at his heart, compelled Talfryn to glance northward. There in the distance, morning's rosy light dusted the Stonefolk Circle. *Which morning? Are we too late?* "We don't need the Horn." He pointed towards the small ring of standing stones. "Father is being held on that hill."

Disa pressed her hand to the base of her throat as she stepped in the direction he indicated. "Do you think Mab is there, too?"

"She won't be far." Vendel slid the blackthorn from his waist belt.

Fenella removed the dowdy brown mantle and replaced it with her glittering emerald fairy cloak. Puffball jumped from Talfryn's pocket to Fenella's shoulder.

Talfryn noted the pixie's partiality.

"Right place. But the question remains," Vendel said. "Whose magic brought us here?"

Sweet scent of apple blossoms threaded the air, obscuring the bitter, metallic scent of Cold Iron. Glimmer materialized afore them and brushed back a ropelike lock of her tawny hair. "Your quest was successful." She spoke as if she had been privy to the outcome. A golden apple branch appeared in her hand. She touched its tip to the glowing serpent bracelet and the wrist cuff vanished.

"Are you here to stop us, Godmother?" Talfryn asked.

"I considered it." Glimmer's shrewd gaze flicked on Vendel afore returning to Talfryn. "But I decided 'twould be more interesting to allow you to proceed…for the moment. As I said afore, I am fond of you and Beacon."

"Then we are not too late?" Talfryn's voice rose with newfound hope.

"Watch with keen eyes. See for yourself if time has run out for one, but mayhap not for another," Glimmer said coyly. "I thought it best to intercept you," she explained, her dragonfly wings thrumming. "The Norse glamour served you well, but 'tis not suitable for the Court."

Talfryn cast an anxious glance at his mother. "Wait!"

With glistening eyes, Glimmer stepped back and flicked her apple wand.

Magic rippled his scalp. His long Norse hair shrank off his shoulders and thickened into short yellow spikes that sprouted like moor grass on his scalp. Enchantments shivered down his neck, then snaked across his shoulders and down his arms. His heavy Norse muscles dissolved in a rip of pain. The magic spiraled down his body, bringing a giddy lightness and the sickly scent of sweet briar ale. The whirling in his stomach stopped. Gone too, were the scratchy Norse clothes. He smoothed his hands over his bark-brown fairy tunic and breeches.

"Já!" Disa outstretched her arms, clearly forgetting herself and intending to close the distance between them and hug him. With an alarmed glance at her weapons, Talfryn warded her off.

Glimmer's pale blue eyes shifted to Disa. "You!" Spying his mother's Cold Iron, Glimmer gasped, then retreated a pace. "Weren't you banished?"

Talfryn stepped protectively between the two. "She's here to return the Aeternitas Potion."

"In person?" Her words rasped as though the ill effects of Cold Iron had settled into her throat. "Very brave. Or very foolish."

Glimmer cleared her throat. Her calculating stare then drifted to Fenella. "Something about you has changed, Thaness." She tapped one of her long fingers against her sparkly lips. "Quite interesting. Let's make sure my dear sister notices." Glimmer flicked her hand. A scoop neck gown of dark forest green replaced the faded léine beneath Fenella's fairy cloak.

Fenella fingered the soft fabric. "'Tis beautiful, dear Royal Sister."

"I haven't finished." Glimmer flicked her hand again. Fenella's wind-tangled hair wove into a thick, elaborate braid threaded with a delicate leafy vine.

Then the fairy godmother leaned close enough to Talfryn for him to smell her gingery breath. "Don't tell our dear Queen I abetted you."

"By my vow," Talfryn answered.

Glimmer gestured towards the distant hill. "Time runs out." With that, she vanished.

Talfryn paced. "Mother, I cannot take you afore Mab. It is too dangerous."

"Since Glimmer knows Disa is here," Fenella argued, "we cannot leave your mother unprotected."

Disa strode as if tugged by an invisible cord to the landward overlook, then stared at the small stone circle in the distance. "If you do not magic me to Beacon, then I will walk to him." She gave her dirt-smudged garments a rueful glance.

Talfryn drew a symbol in the air, then flicked the magic. Her shield-maiden's padded tunic and leggings appeared cleaner, newer, and brighter.

"Thank you, son."

"What aboot her weapons?" Fenella asked.

"My sword is dwarf forged," Disa said, protectively clasping the hilt. "I am loath to part with it."

"Leave your Cold Iron here," Fenella directed, moving to the shield-maiden's side. "I have an idea, but its only hope for success is if you enter the Court unarmed. Vendel can retrieve your weapons later, and Talfryn will find a way to return them to you."

Frowning, Disa traced her thumb over the iron fretwork surrounding the inlaid moonstone. "This sword has been my constant companion. Without it and my shield, the Fates would have taken me long ago." Her eyes glistened as she focused on Fenella. "The druid foretold your presence, that I would find allegiance with you. Since Cold Iron cannot be concealed in Fairy, 'twould be foolish to arrive armed when I present myself afore Mab." She paused, inhaled deeply, then exhaled a long, decisive breath. "I shall place my faith in you, Thaness, and I pledge you my fealty." She removed her shield and sword, then stowed them between boulders in the rock outcrop overgrown with heather.

"I thank you," Fenella said.

Disa nodded, then bent and rubbed wild thyme between her palms. "To remove the smell of Cold Iron," she explained.

"Clever." Vendel cleared his throat, then turned to Talfryn. Dirt and smoke clung to his skin and clothing. "I am hardly presentable."

"Can we trust you?" Disa glared down at him.

"Aye. While I had hoped to succeed after so many afore me had failed, I never wished Beacon harm. I yield to Fate."

Talfryn worked his magic. The dwarf glanced down at his freshly cleaned skin and tunic. "I thank ye."

"You're welcome." Talfryn turned to Fenella. "What is your desire?"

"Take me to your queen."

"'Twill be dangerous."

Fenella straightened her spine. "I am ready." She offered her hand. "Let us save your father. Fate wills it."

Taking Disa's hand and Fenella's, Talfryn expelled a long, worried breath. Puffball leapt from Fenella to Talfryn's shoulder, then scrambled inside his cloak pocket. Vendel completed the circle. Ill to his stomach, Talfryn said, "The Stonefolk—"

Fenella squeezed his hand.

Courage flooded his veins. Fenella's thistle-green eyes reflected confidence and support. Sea air filled Talfryn's lungs, sending strength to his muscles and bones, vanquishing any lingering effects of the Cold Iron. He squared his shoulders and stood straighter.

"The Stonefolk Circle."

CHAPTER FORTY-EIGHT
THE MEETING

The Stonefolk Circle
The Realm of Earth, Fairy

Landing outside the stone circle, Fenella choked on air deadened by suffering and desperation. She sensed the spirits of the standing stones had withdrawn within themselves, as though they wanted nay part in external happenings.

Disa reached for her sword. Then, apparently realizing she had left the weapon on the headland, the shield-maiden lowered her arm to her side whilst still surveying the area, her eyes keen and focused.

Talfryn ran towards a dying tree.

"'Tis a rowan!" Fenella warned, confused by his action. Surely, he knew what kind of tree it was. "Beware!"

The Halfling halted, then slid his foot forward another step, as though he were torn whether to approach or not, resisting some desperate urgency to proceed.

A nearly inaudible moan from the far side of the tree reached Fenella's ears. "Beacon," she said under her breath. Her stomach whirled from the lingering presence of dark forces and the specter of death, but she forced herself forward.

Spying Beacon curled on the ground with his knees near his chin, her blood iced. His once vivid butterfly wings, still attached to his back but flattened together, appeared tattered and soiled. A heavy rope, linked with Cold Iron to a glowing cuff on Beacon's wrist, tethered him to the rowan. *Rowan and Cold Iron? Is Talfryn's father dead?*

"Beacon!" Disa dashed towards her husband, but Talfryn caught her around the waist and restrained her.

Beacon's eyes flickered, then closed, as though the effort had exhausted him.

"He's dying!" Disa struggled to break free.

"Beware." Talfryn strained to hold his mother. "Mab!"

"Back away!" Fenella cried as she and Vendel rushed to pull mother and son from the rowan and Cold Iron.

Of a sudden, the air afore them glinted. Gold fairy dust flurried, carrying exotic, spicy scents. The flurries gathered into a hot whirlwind that ripped lousewort and faded violets from the ground. Fenella clutched her wildly flapping cloak and held down her gown.

Two imposing fairies with glittering dragonfly wings materialized within the tempest afore the swirling winds calmed. Fenella's gaze brushed over the male. He stared—nay, he appraised her—with disquieting curiosity.

Talfryn leaned close and whispered in her ear, "Brindleburr, the Prince Consort."

Fenella forced her features into an unimpressed mask. *I wager you have lured many a human, and fairy, too, with your handsome glamour. But you willna entice me.* Unexpected hints of gray at Brindleburr's temples drew her attention.

Noticing her stare, the fairy quickly conjured an oversized, foppish, felt hat. He planted the creation low on his head, folding the sides up like a sable swan's wings. The style clashed with the expensive cut of his black tunic and the dark purple trousers that billowed loosely over his thighs afore fitting closely at his ankles. By magic or trick of the light, when he moved, the trousers changed color and appeared dove gray. His impressive gold wings, edged in a bold raven-black stripe with a soft white rim, twitched twice.

Power pulsed from the regal female beside him. Her deceptively simple gold crown, adorned at its center with a golden dragonfly flanked on both sides by three amethysts, drew attention to the fairy's imposing height. A pleated, gold gossamer gown clung to her long, lean body, revealing her bronze shoulders and arms.

Mab.

Three children, two dark-haired lads and a redheaded lass, each wearing a glowing spell ring around their head, peeked out from behind the royal couple. They appeared to be seven to nine years old. Waiting like obedient dogs, they remained on their hands and knees behind Mab.

Talfryn whispered in Disa's ear. She nodded, her face a murderous mask. He released his hold around her waist and moved aside.

An emaciated, grim-faced dwarf bearing a woven sack materialized next to Beacon's prone form. The contents of his bag clinked together as he swung the sack off his shoulder and placed it on the ground near Beacon's feet. This once strong, vibrant Trooping Fairy twitched, then moaned as though anticipating great pain. The dwarf hunched his bony shoulders as if in empathy and knotted his smudged hands together.

Clasping his mother's wrist, Talfryn whispered, loud enough for Fenella to hear, "Await the queen's invitation."

The dwarf with the sack must have overheard, for he glanced their way. His jaw dropped slightly and his eyes, dull, Fenella imagined, from a life of harsh servitude and neglect, flashed briefly with hope afore he glanced at Mab and shuttered his expression.

Mab brushed from her face a few narrow green braids and glittering gold strands that threaded her long, bark-brown hair. When she raised her gaze and peered at them with eyes like white eggs, Fenella hid a gasp behind her hand.

"Dear Queen," Talfryn said. "May I present Fenella, Thaness of Thorburn. Fenella, this is Queen Mab."

The redheaded slave girl crawled around beside Brindleburr and rose to her knees. The lass's intent, almost hopeful stare arrowed onto Fenella, momentarily distracting her. But the Prince Consort flicked his hand, and the keen-eyed child once again dropped to her hands.

Fenella directed her attention back to the queen, then bent in a quick curtsey.

Mab leaned forward, then in a burst of breath, exclaimed, "Your hair!" She shrank back. "Brindleburr, doesn't the color remind you of—"

Disa shifted from one foot to another, attracting Mab's peculiar gaze; small, pale blue dots in the queen's unsettling eyes darkened like the midnight sky.

"You!" Mab's oversized dragonfly wings fluttered. Whiffs of frankincense, bitter orange, and cinnamon scented the air. "Halfling, how dare you bring that banished human to my Court!"

"She wishes to return the Aeternitas Potion to you…in person." Talfryn explained.

"Disa is here under my protection," Fenella said, her voice commanding but respectful, "as an honorary member of my clan."

Mab glanced from Fenella to Disa, then whirled. "Brindleburr."

The Prince Consort flinched. "Aye, dear Queen?"

"Has this ever happened afore? A banished human allowed to return?"

"I do not think so, dear Queen. But we do have a pact with the Thorburn clan, and if Beacon's wife is here under the thaness's protection—"

Mab held up her elegant hand. Brindleburr swallowed his unspoken words and stepped back.

"Bring me the Aeternitas Potion," Mab commanded Talfryn.

Disa's hands trembled. She lifted the hem of her cloak, then picked at the seam with her fingernail.

Mab crossed her arms over her chest and huffed impatiently.

Talfryn knelt afore his mother with his back to Mab. He slipped the claw-shaped thunderstone from beneath his tunic, then used the needle-sharp nail to shred the stitches. Disa held the ampule in place while Talfryn tucked the glowing symbol of the Noble House of Storms beneath his neckline.

Fenella held her breath as Talfryn extracted a narrow, smoky crystal vial. His fingers reddened as though heat soaked into his skin. Rising to his feet, he whispered, "Everlasting life."

An insistent pulse throbbed in her chest above her heart, and a bloom of warmth spread like a flower unfolding. She glanced at Vendel. The dwarf stared at the vial with unwavering intensity while rolling the Blackwood staff back and forth between his palms.

Talfryn faced Mab and Brindleburr. "Respectfully, dear Queen, afore I approach, I request the Cold Iron be detached from my father."

"Beacon is in nay condition to flee should the potion prove false," Fenella added. She swallowed hard, suddenly parched, as Mab's inscrutable stare burned into her.

A moment of silence passed, filled only by Beacon's labored breaths. Then Mab gestured, and the wrist cuff and Cold Iron links vanished. The dwarf's narrow shoulders slumped with apparent relief. He then gathered up his sack and scurried back several steps.

"Thank you, dear Queen." In three strides, Talfryn reached the monarch. With a slight bow, he held the vial aloft.

Mab's gold-and-dark-green lips pursed as if a sour memory had struck her. The queen glanced about, appearing almost ill at ease, afore she extended her arm and her long fingers closed like tentacles around the vial. A tear escaped down Mab's cheek as she held the vessel close to her eyes. She studied the crystal ampule for a moment, then weighed the vial in her hand.

"It appears none of the potion is missing." Mab handed the vial to the Prince Consort as if she held a dead rat by the tail.

Tight-lipped, Brindleburr conjured a gilded box that resembled a miniature bishop's coffin. Instead of a holy man's effigy, the lid depicted a gold creature with the body of a lion and the head of man. The curious box floated in the air. Its lid separated from the bottom, revealing a tiny, indented pillow inside with small drawings of birds, human body parts, and curious symbols painted on the stiff cloth. Brindleburr nestled the vial on the cushion, then returned the lid to the box. Magically, the latches on either side of the shining container clicked closed.

The queen waved her hand, and the gleaming chest vanished.

On the ground, Beacon stirred.

Mab's gaze fixed on the frail dwarf who had carried out her

bidding. "Say nothing of what you've seen here Bleak, or I will sever your tongue."

Bleak bowed his head. "There is nothing to speak of, Your Majesty," he said in a tremulous voice. His knuckles bleached white as he clutched the sack.

"Good." Mab flicked her hand, and Bleak disappeared. Her pleated gown rustled as she turned to the Prince Consort. "I have an announcement to make. Summon my subjects to the Great Chamber."

"As you wish, my Queen." Brindleburr, his flawless skin now almost as ashen as Beacon's, bowed. Apparently forgetting he wore a hat, he caught the creation as it fell. His wings fluttered nervously as he vanished in a swirl of black and sunflower smoke.

"May I offer a few words?" Fenella asked the queen.

"Not 'til we have assembled under the hawthorn." Mab's ghostly gaze swept Fenella's fellow travelers. "The stone circle is a portal. Gather Beacon, enter its ring, and I will summon you momentarily." Mab vanished in a twist of blinding wind, leaving behind glittering gold dust to dissipate as it fell to the ground.

Chapter Forty-Nine
Flee or Fight

Afore the last fleck of gold vanished, Disa reached Beacon. Fenella's heart ached for her. The shield-maiden's tears splashed her husband's wan face as she kissed his cheeks, then cradled his head. "Magic him to some place safe," she begged.

"Is there such a place?" Puffball flattened herself against Beacon's chest. "Talfryn, perform a Restore spell."

"Nay spell has ever countered Cold Iron poisoning. Mab tests ye. Tests all of us. She expects ye to flee," Vendel said, "and to cover up her cruelty by attempting a Restore spell."

"I have to try. He is near death."

"Attempt the spell and ye will face Mab in a highly weakened state."

Talfryn moved toe-to-toe with the dwarf. "I noticed the look that passed between you and my father when we first walked to the Stonefolk Circle. Whatever pact you two made, will not apply. If Father dies, it will be at the hands of the dwarf who administered the Cold Iron, not Mab. Besides, now she has the Aeternitas Potion. What need has she of us?"

Fenella placed a gentle hand on Talfryn's arm. "If you flee, Mab will be unchallenged. She will turn everyone against your parents. The fairies will believe whatever story she concocts. And if Beacon dies, she will swear *he* confessed to killing Merrybelle. Mayhap, she is telling the fairies that as we speak."

"Fairies cannot lie," Talfryn pointed out.

"Aye, but Mab can tell a partial truth that will lead others to conclude something Mab knows is false," Fenella countered.

Disa stilled. "The soothsayer told me Mab fears the truth above all else."

"Ye know what truths she is trying to conceal." Vendel shook the blackthorn staff. "Who killed the Last Fairy King and Gudbrand?"

"Who killed Merrybelle?" Talfryn added.

"And why is an immortal fairy's hair turning gray?" Fenella asked.

"What?" Talfryn and Vendel said at once.

Fenella rose to her feet. "Quick! Talfryn, magic us, all of us, to the hawthorn Mab mentioned. We will use Mab's worst fear against her."

"But Father—"

"I believe he would want this."

"But Mab told us to enter the circle," Puffball pointed out.

"Which could be a portal to anywhere," Fenella argued, revising her plan in her head. "Why do you think Mab dinna threaten to cut anyone's tongue except Bleak's?"

Vendel touched his lips, then stroked his beard. "Mab doesn't fear we will expose her, because once we enter the circle, we will vanish…forever. 'Tis a trap! She will send us to a realm where our doom would be all but certain, where dying would be *our* failing."

Fenella nodded. "Nay witnesses."

"Merlin's beard," Talfryn swore.

"We must expose Mab's cruelty to Beacon," Fenella said. "And what I suspect Brindleburr did to Merrybelle."

"Brindleburr?" Talfryn's jaw dropped.

"I thought he liked Merrybelle," Puffball said.

Vendel crinkled his brow. "Mayhap 'tis the very reason. If he was too fond of Merrybelle…"

Talfryn paled. "And provoked Mab's jealousy—"

"Mab would have forced the deed upon him." Vendel scowled. "I should have guessed."

"Aye," Beacon mumbled, his eyes sweeping once beneath his closed lids.

"If Brindleburr is already aging, the potion willna save him." Fenella's pulse thundered. "Help Beacon to his feet," she directed Talfryn and Disa. "Puffball, gather flower dew. Remember how it strengthened Talfryn when he suffered from Cold Iron poisoning."

Puffball slipped off her skullcap and darted to the lousewort.

"Father, can you retract your wings?"

Beacon's mouth remained slack.

"Stay back," Talfryn warned Fenella, "lest the Cold Iron in your boot harms him beyond repair."

Fenella retreated and drew the dirk from her boot. But she couldn't bring herself to part with the weapon, the only relic of her paternal lineage. *If the dirk were nae my late father's, I would throw it as far from Beacon as possible.* Resolved to keep her distance from Talfryn's ailing father, she stowed the knife back inside her boot.

With great care, Talfryn and Disa lifted Beacon to his feet. Flanking him, each with an arm around his waist, they held him upright. His head slumped forward.

"Hold his head up," Puffball said as she scampered up Tansy to perch on his shoulder. With one spindly hand under Beacon's chin and the other holding her brimming cap, the pixie poured flower nectar and dew into Beacon's mouth. Beacon gulped, then coughed.

"Thank you, Puffball," Fenella said.

The pixie nodded, then squirmed into Talfryn's cloak pocket.

"Form a line, not a circle," Talfryn instructed.

Fenella frowned, uncertain how the magic would work. Still, she positioned herself behind Vendel, who appeared equally nervous and unsure.

Talfryn said, "The Great Chamber, beneath the hawthorn."

Then Fenella flew—*flew by herself*—like a well-aimed arrow towards a knot in a hawthorn trunk. The air afore the knot whirled wider and wider, pulling her into the tree. Magic whisked her down through the hawthorn and below its roots. Her feet hit tamped ground with a painless thud.

A floating cut crystal orb skimmed the top of her head. It cast an amber light on the fay, dwarfs, and some odd beings Fenella could nae identify. Root tangles braced the earthen walls and curved cavern ceiling. Flower scents, loamy forest smells, acrid smoke, and inexplicably a damp dog odor, assaulted her.

A lifetime of Gran's warnings screeched in Fenella's ears. *"The fairies stole my sister, and Esme was never seen again! Never enter a fairy ring. Never eat fairy food or drink fairy drink. If you do, you'll be trapped in Fairy forever."*

Instinctively, she raised her hands to waist level. The center of her palms pulsed. Then to her great relief, Vendel joined her. She followed his glance, spotted Disa, and then Talfryn carrying Beacon.

A sudden awareness tugged Fenella's attention towards a fay standing in front of the shadowy back wall. Not recognizing the squat, mottle-skinned fairy, Fenella averted her gaze. But a prickle down her neck sent her swiveling back. The fay raised an apple branch wand, just for a heartbeat, then inclined her head slightly as she lowered her hand. Their gazes locked.

Glimmer!

The glamoured Royal Sister pressed her forefinger to her lips, her expression now hardened in warning. Unnerved, Fenella nodded once, then returned her attention to the crowd.

A collective gasp rose from the fairies standing wing-to-wing inside the chamber. Those closest to Fenella surged forward.

Show nay fear, Fenella told herself.

"Is it true? Beacon of the Noble House of Storms is dead?" shouted a male fairy whose elegant garments and proud posture indicated he, too, was of a noble house.

"First Merrybelle, now Beacon?" came another outraged shout.

Diminutive goblins clicked their pointed teeth.

"My father is not dead yet, though our dear Queen ordered him shackled to rowan with Cold Iron." Oddly, Talfryn's tone sounded non-judgmental.

Does he bide his time 'til I can reveal my plan? Or does he contrive one of his own?

Gasps, like a contagion, spread through the large chamber. The sudden fluttering of wings stirred the air.

"But there is nay proof Beacon stole the potion!" Protests buzzed through the gathering.

A female fay with long lavender hair shouldered through the throng. As she neared, Fenella noticed oozing welts on her pale skin and her tattered wings. *She is the equistragon rider whose whirlwinds saved us.* The fairy edged near Talfryn. Slipping her elegant hand on his shoulder, she pressed close and whispered in his ear. Her scorched wings fluttered, stirring the cloying air.

"Her name is Lily," Vendel whispered.

With a pang, Fenella watched as Talfryn's eyes lit up with relief, gratitude, and mayhap, something more. *Is Lily the true reason I did nae see Talfryn 'til he needed my help?*

Chimes whistled sharply. The fairies retreated, muttering in shocked tones. Those with wings fluttered them. Some stared at Beacon and wept. Lily squeezed Talfryn's arm, then moved on to the front of the gathering. Her tearful glance alighted briefly on Fenella afore sweeping to a gold runner on the ground.

Fenella followed Lily's gaze to a raised dais, where Mab and Brindleburr loomed over the assembly. Mab, her long-fingered hand fanned against her breastbone, gaped at Talfryn and Beacon.

Mab dinna expect to see them, or any of us, again. Ever.

The Fairy Queen must have sensed that every eye was upon her. She nudged Brindleburr. As the Great Chamber quieted, the royal couple stood aside as the three child slaves from the stone circle swarmed the thrones and swept them with blooming lavender stalks. The children shrank back once they completed their task, and flower fairies flew in to drape dark purple silk over the thrones.

While the queen and her consort sat down, Fenella studied the children. *None of them have tilted eyes, pointed ears, or wings.* She glanced around. In the teeming cavern, none of the others wore glowing circlets.

None of the others are spellbound. Or children. Or save for Disa, the dwarfs, and me...human.

Fear chilled her blood. Another of Gran's warnings rushed back to her. *Fall down a fairy mound, and the fairies will enslave you.*

"Halfling!" Mab's shriek pulled Fenella back to the present. The fairy queen's unsettling stare was locked onto Talfryn as she compressed her lips into a tight line.

"You summoned us, dear Queen?" Talfryn asked guilefully.

Mab opened her mouth to retort, but then closed it as she apparently assessed the hostile mood of the crowd. She rose again, rustling her pleated gown. Ignoring Talfryn, she directed her words to the assembled fairies, dwarfs, goblins, and brownies.

"When I last saw Beacon of the Noble House of Storms, he was very much alive," Mab claimed. "And as I was about to say afore…summoning Beacon's Halfling son, Beacon's human wife has confessed to stealing the Aeternitas Potion."

Murmurs rippled through the crowd.

Noting the satisfied twist to Mab's lips, Fenella signaled Talfryn to lower his father to the tamped earth. Then she approached the gold runner. "Permission to speak, Your Majesty?"

The queen raised a glittering brow. Then, as though aware of the palpable anger in the chamber, she said, "Permission granted."

"For those of you who do nae know me, I am the Thaness of Thorburn," she told the crowd afore directing a reasonable tone at Mab. "As you have already rightly determined, Beacon and Disa did nae steal any of the actual potion," Fenella said. "Indeed, Disa, unarmed, bravely and respectfully returned the vial in person, the moment you discovered it missing. So nay real harm was done. Clearly, Disa bore nay ill intent and is sorry to have inconvenienced you."

"Inconvenienced me?" Mab stared at Fenella as if she had grown a second head. "She *stole* from me!"

"In the human world, parents will commit desperate acts to be reunited with their children, especially their newly born." The red-haired slave lass peeked around the dais. Distracted, Fenella lost her thoughts, retrieved them, then continued. "I ask only that you permit Disa to live on Thorburn land unharmed. She will never return to Fairy again without your permission."

"Dear Queen—" Beacon's rasped words died on his lips. He closed his eyes once more. Seeing his chest rise and fall, the knot in Fenella's stomach loosened.

Mab's steepled hands grew rigid as she thrummed her wings. "Dwarf. Halfling. Help Beacon to a sitting position so I may better see his face," Mab said at last, stilling her wings.

The crowd parted. As Vendel strode to Talfryn's side, the assembly held its collective breath.

Joining Fenella, Disa watched her half dead husband eased into a sitting position, then shrank against Fenella and whispered, "We are in grave peril."

Fenella risked a quick glance at Glimmer. *Mayhap I can appeal to her.*

The disguised Royal Sister, her intent gaze trained on Talfryn, Beacon, and Vendel, shifted her focus to Mab.

"I disagree with the thaness." Mab paused, seemingly for effect. "*Real* harm has been done." She lifted off the dais then, with deliberation and menace, glided half way down the gold path.

Fenella and Disa retreated to side with Beacon, Vendel, and Talfryn.

"A crime was committed. Against this Court," Mab declared, landing silently on the gold runner as Brindleburr flitted from the dais to stand beside her. "Beacon and his human wife do not deny their guilt. Nor do they show remorse." Mab glared at Disa. "Therefore, a measured and harsh consequence must be exacted so that no other dare act against me. I rule...one of them must die."

Beacon jerked as though slapped. Disa bowed her head.

Sensing the royal fay's strong affinity for drama and need for spectacle, Fenella acted swiftly. "Again, I would speak, Your Majesty?"

Seeming somewhat perturbed at the interruption, Mab nevertheless granted permission with a flick of her hand.

"Dear Queen, in a masterful way, you have already punished Beacon and Disa quite severely."

Mab angled her body towards Fenella as though surprised, yet curious to hear more.

"By nae calling for the potion," Fenella continued, "you have caused Beacon and Disa years of torment as they waited for the moment of reckoning. For eighteen years—their son's entire lifetime—you have tortured Beacon and Disa by separating them, and by keeping Disa from her son. Talfryn has grown up without her. There can be nay greater heartbreak for a human mother than to be denied her child. So, Queen Mab, in fact, you have already exacted the agonizing punishment you require."

All was silent, save for Beacon's ragged breaths. Fenella clutched the edge of her cloak, awaiting the queen's response to her interpretation of the events.

Mab's lips quirked upwards in a vicious smile. "It *was* very clever of me." She appeared quite pleased with herself. "Aye, a fate worse than death." Seemingly mollified, she flicked her hand with a complacency only a queen would brandish.

"Arise Beacon, Head of the Noble House of Storms, and stand beside your wife," Mab commanded.

Fenella sensed Talfryn's heart hammering as he and Vendel assisted Beacon to his feet. Disa took Vendel's place and slipped her arm around Beacon's waist. A crystal lantern floated above them. Its unforgiving light revealed the age lines in Disa's weather-beaten face. Beacon's head drooped towards his chin; his once lustrous snowy curls, now gray, were matted to his skull.

"I have decided to grant your wish, Thaness," Mab said, her

tone regal. “Disa, the banished human, will be allowed to live out her natural life on your lands, unharmed by the Blessed Court.”

“Esteemed Queen.” Fenella concealed her astonishment with a deep bow. She couldna be sure, given the queen’s strange eyes, but she thought Mab skewered Beacon and Disa with a cunning look.

“Beacon, should you recover, you may visit your wife on Thorburn land whenever your duties permit.”

“Thank you, dear Queen,” Talfryn spoke on his father’s behalf. “With your permission, may we take our leave? My father requires urgent attention.”

“Permission granted. Though, sadly, there is nay known cure for Cold Iron poisoning, should that be what ails him.”

As a low mumbling rippled through the crowd, Talfryn raised his voice to be heard. “I thank you, dear Queen.” There was silence, then he asked, “Perchance, while I was away, was Merrybelle’s murderer discovered?”

Mab glanced at Brindleburr. He averted his eyes. Addressing the crowd, she announced, “The investigation continues.” Then she set her guarded glare on Talfryn. “See to your father. Return the humans to Thorburn land afore nightfall.”

Talfryn bowed. “As you wish, dear—”

The queen vanished in a flurry of gold dust that blew the hat off Brindleburr’s head. Startled and abandoned, the Prince Consort dove for his hat afore he, too, disappeared.

As Disa helped Talfryn gather up Beacon in his arms, the disguised Royal Sister appeared at Fenella’s side and whispered in her ear.

“Alliances are being tested, Thaness. If a fairy kills a fairy, he or she will lose their immortality. But if a human kills a fay, mayhap a fay many in Fairy believe should step down… or better, disappear…” Glimmer arched her brows.

Afore Fenella could retort, Glimmer melted into the dispersing crowd with fairy speed, then vanished in a whirl of mottled apple blossoms.

Chapter Fifty
Two Paths

The Blackwood Stronghold

All that was left of Edward's betrothed were footprints in the dirt and the haunting echo of her scream. Dread choked the breath from his lungs. Gaping at the spot where Fenella had stood, Edward squeezed his eyes shut, then opened them.

"Did you see her? I thought I heard…"

Calidore. Edward spun towards the voice. "How did she flee past you?"

"I was nae guarding her door. His lordship dismissed me for the night. Derry received similar orders. Lord Argonshire said Nurse would watch while the thaness slept."

Scowling, Edward shoved the tangle of keys against Calidore's stomach; the guard took them without question. "Who raised the alarm?"

"I doona know."

"Where is Derry?"

"The maid wanted to check on her lady's horse. She should have run right past you."

"Och, I was too preoccupied to notice." *Though I canna see how I could have missed her.* "I will search the stables. Find Connell

Thorburn. Inform him I wish to speak with him. Tell my men to watch for the MacLeod and the dwarf."

"As you wish."

Edward rushed to the byre, aware guards thudded down the donjon's outer stairs. Within a moment, they would storm the upper courtyard, and some would dash to the stables. He needed to arrive first; see if indeed, Derry was there. *Why would a lass who avoids all horses, save the white mare, run to the byre?*

Fading daylight slanted through the open stable door despite the late evening hour. Edward entered and inhaled the smells of summer hay, horse, and leather. The commotion outside had set the horses on edge. As he led with his sword, palfreys paced and whinnied on either side of him. A sable mare Fenella usually rode bobbed her head as he passed.

"Starlight." Derry's voice rose childlike from the rear of the stable. "I doona believe your true mistress is from the village. And I doona know why she bestowed you upon my lady. But the bailey reeks of potent, vengeful magic. My lady is nae safe here, or wherever she was taken. So, I beg your mistress to protect Lady Fenella if she is able."

Vengeful magic? A rival beyond my reach. Edward stepped closer, which caused the white mare to whinny and back away.

Derry spun around, her hand to her mouth. "Mr. Blackwood!" Afore she could say more, Angus dashed in, his sword raised. He halted when he spotted them.

"I have already searched." Edward sheathed his sword and glared expectantly. Angus hesitated, then withdrew his weapon. "Tell the guards to raise the portcullis," Edward ordered. "Search the lower courtyard."

"But how could she—"

"Just do it!"

"Aye, Mr. Blackwood." Angus retreated, quickening his pace as he left.

"Follow me," Edward told Derry.

The maid complied. Edward wondered anew why her gait resembled a drunken seafarer's, or why it often became normal once they were outdoors. As men and hounds swarmed the courtyard, Edward led Derry behind a towering beech at the side of the byre.

"Where did you smell the vengeful magic?"

Derry's eyes widened. She glanced away as though seeking escape.

"Lady Fenella needs our help, Derry."

She turned her face towards him. Her voice trembled. "Near the tree where you were standing earlier."

So, you did flee past me. "Good."

"Good?" Her fingers pressed her lower lip. "You believe me?"

"Aye."

A hound barked. Derry visibly startled and placed a steadying hand against the tree as three men on horseback rode past. Edward waited 'til they had reached the raised portcullis.

"If you witnessed someone stolen by magic—vanished afore your eyes—how would you find them?" Edward asked.

"I am nae sure. Wouldna you need to fight magic with magic?"

"Is that why you sought out the gift horse?"

Derry nodded. She dropped her voice to a whisper. "I believe 'tis a fairy horse."

Fairy horses. Magic. People vanishing. Edward tucked a stray lock of hair behind his ear, then said, "I hope the mare's mistress can help Lady Fenella."

"Aye, though we doona know the mistress's intent. Does she wish to protect my lady, or harm her?"

"Let's hope she wishes to protect Lady Fenella."

The maid's brows drew together, suggesting she had doubts. "'Twould be better if we had our own magic."

Indeed. He thought of the Romani pouch hidden inside his pocket. "I will find Lady Fenella, and keep her safe," Edward promised. "I thank you, for your help."

Derry bobbed an awkward curtsey.

"You may take your leave."

Without further ado, Derry fled towards the garden.

Edward scanned the upper courtyard. Then like a thief, he slid around the corner and slipped through the stable door. Already, several horses and their tack had been taken. Straw crunched beneath his boots as he conducted a quick search. Seeing he was alone, he drew alongside his horse. The black nuzzled his shoulder as Edward pulled the linen pouch from his pocket and loosened the strings. He hesitated, rubbing his fingers across the yellow cloth as he remembered the old woman's warning.

Your palm shows two possible destinies. First path: return home and trust in the thaness. Have faith she will find you. Or second path: employ magic, and risk changing both of your destinies forever.

He swallowed hard. "I trust Fenella," he said, hoping to ward off ill omens. Remembering the fear in Fenella's voice afore the MacLeod, or whoever he was, spirited her away, Edward set his jaw, then upended the small sack. A ring fell into his palm. Its black stone pulsed when he held it up to the light. Colors and shapes appeared on the smooth surface. His horse whinnied and backed away. Dazzled, Edward drew the ring closer. Thorburn's fosterling appeared in the stone. With horror, Edward saw the blond's true form. As the scene magically widened, his muscles clenched.

The rogue has taken her to what...must be...Fairy.

His resolve hardened, Edward slid the heavy ring onto his finger. The black reared. Distracted, Edward paid nay heed to the tingling in his palm as the line representing the first path faded from his hand.

Chapter Fifty-One
Death's Door

Belle Flor, The Noble House of Storms
The Realm of Air, Fairy

Fenella rushed to Beacon's chamber, spilling precious drops of the hot curative tea she carried. "Pox!"

Talfryn met her in the doorway. "The Restore spell didn't work. Father worsens. And I have been unable to reopen the secret tunnel."

"Did you wash his wrist and arm where he was cuffed and apply the poultice I concocted?"

"Aye. The cleansing and dressing relieved the burn from the Cold Iron cuff."

"Good. Why doona the healer arrive in her human form without using the tunnel?"

Talfryn lowered his voice to a faint whisper, forcing Fenella to lean close. "Mab has nay knowledge of Vendel's mate. I vowed to the dwarf I would safeguard Hilde's secret."

"I see." Fenella pushed the steaming bowl into Talfryn's hands. "I used chamomile, dittany, and a tiny amount of another plant whose name escapes me. The combination should draw out the poison and help break Beacon's fever. In proper proportions, the cure works on humans..."

"But may fail with fairies."

"I fear so. 'Tis impossible to predict." She glanced past him to the open circular window. Her heart jumped at the fading daylight. Time was a plague-infested hound chasing her heels. If she dinna find a way to stay ahead of the beast, Beacon would die, and her destiny and Disa's would be in Mab's merciless, long-fingered hands.

I'll nae allow Mab to magic a slave spell ring on me!

With deliberate steps, Talfryn walked the brimming bowl to Beacon's floating bed. "Let me cool it a little first," he warned Disa as she reached for the burning hot drink. He placed the bowl on a quickly conjured bedside table. Murmuring a command in a strange lilting language, he passed his hand once over the vessel. The aromatic steam rising from the tea lessened.

Fenella lingered in the antechamber, loath to remove her hidden, jewel-encrusted dirk. Mayhap Belle Flor was magically warded against thieves. Still... Nay locks or guards were to be seen, and fay folk are attracted to beautiful, shiny objects. She would nae risk losing her heirloom to a quick-fingered pixie with a high resistance to Cold Iron. "Use dipping cloths," she instructed from the doorway. "Otherwise, he might choke."

Scallop-edged cloths magically appeared beside the tea. Disa set herself to the task.

Talfryn returned to Fenella's side. He tore his stare from his parents and said, "Your herbal knowledge has grown."

"I had some understanding afore, from my gran. Edward Blackwood taught me much more. We have been restoring his brother's herb garden." An ache clouded her heart.

Talfryn touched her arm. "You will be home afore dark."

"Nae to Edward. 'Tis too dangerous. Besides, we must take Disa to Thorburn land. Uncle Connell's stronghold would be the safest. My other uncles are too… Uh... Connell will protect her."

Talfryn held her gaze, stuttering her heart. He parted his lips to speak, but she cut him off. "Time runs out," she reminded him.

"Aye." For a heartbeat, his beguiling fay features—tilted eyes, high cheekbones, lush lips, and slightly tilted nose—scrunched into a very human expression of confusion and regret.

He wiped his hands on his tunic. "I'll try the Unsealing spell again."

At Beacon's bedside, Disa dipped a cloth into the tea, then dripped the drink into Beacon's mouth.

The fairy coughed after several drops. His eyes briefly opened, leapfrogging Fenella's heart and nay doubt, Disa's. The next dribbles safely coursed down his throat.

A tepid breeze from the maze garden eased in through the circular window. The draft, stirring sea glass wind chimes, carried the sweet fragrance of dog roses. *And still*, Fenella thought, *Beacon burns with fever and reeks of sweat and poison. We need Hilde.*

A thought came to her. "Remember when we saw Belle Flor in Stanka's fire? The window was covered."

"An Illusion spell," Beacon mumbled.

"Protection against prying eyes and spies," Puffball said from the bed.

Talfryn plowed his fingers through his spiky hair. "I'll try, but if the spell works, we won't be able to see out."

"And we will nae know when night has fallen," Fenella realized.

"Mab decreed we leave Fairy by nightfall." Fear tinged Disa's voice.

Pox. "Cast the Illusion spell. But make haste," Fenella decided. She glanced at Beacon. "Disa, nay more tea."

"Are you certain?" Disa asked.

"Aye. Watch his fever. It may spike, then lessen as he sweats out the poison. We canna let his fever grow too fierce, or it will damage his insides."

Disa placed the cloth beside the bowl, then clasped Beacon's hand.

The Illusion spell manifested as a heavy black cloth, like a pirate ship's sail. The glamour shrouded the circular window as well as its frame inlaid with enamel birds, butterflies, and flowers.

"Anyone peering in will see only Father asleep in his bed," Talfryn said.

Fenella half closed the chamber door, then stood blocking the open half even though there were no windows in the adjoining chamber.

Talfryn again recited the Unsealing spell.

The air rippled. Fenella held her breath as the mouth of the tunnel appeared. Then she heard a distant, subterranean howl.

"You did it!" Puffball rose from her vigil at Beacon's feet.

"By now, Vendel will have told Hilde that Mab has the Aeternitas potion," Fenella said.

"The news willna stop her. Hilde is a midwife and healer. She will not abandon a patient in need." Disa's fierce expression promised severe consequences should Hilde choose otherwise.

Beacon curled up on his side with a whimper.

"Wash away the poison as it beads on his skin." Fenella returned her attention to the tunnel.

Talfryn quickly conjured a bowl of fragrant rosemary water and a stack of thick cloths and placed them upon the small table.

Disa touched his arm. "I thank you." Her hands trembled as she folded back Beacon's bedding.

"The tunnel!" Fenella warned.

Talfryn rushed to the opening.

A furry head appeared. The graying tawny-eyed wolf paused and assessed Talfryn. Fenella held her breath when the beast's intense gaze jumped to her. Then with a sniff, the wolf dropped the bundle it carried in its mouth and turned towards Disa.

"Hilde!" Disa cried.

With a yelp and kicking up of dirt from the tunnel, the wolf transformed.

Disa's shoulders sagged. "I guarded the vial with my life."

"As ye promised." Hilde swept her bundle off the ground, snapped the dirt from the mantle, then wrapped it around her body. "Vendel told me ye safeguarded the potion to the very end. He will be along later. He needed time alone to commune with his ancestors, explain what transpired and ask for guidance." Hilde approached the floating bed and crinkled her nose. "How is it Beacon still lives?"

"We have given him a tea to expel the Cold Iron poison," Fenella said.

"Indeed! Ye must be the thaness Vendel told me about. And ye!" Hilde clasped Talfryn's wrists. "Ye have grown into quite the youth!"

High color bloomed on Talfryn's chiseled cheeks. "Thank you for coming to Father's aid." He waved his hand, and the bed silently floated to the floor.

Hilde peered at Beacon, then pressed the back of her hand to his forehead. "The tea is quick-acting. His tunic is soaked with sweat."

"I was about to remove it and bathe him," Disa said.

"We must make haste. He is beginning to shiver."

From the doorway, Fenella felt the heat radiating off Beacon.

"His fever has dropped, but it is still dangerously high," Hilde warned.

"I can make a poultice," Fenella said.

Talfryn joined her in the doorway. "We need something quicker. Night will be upon us within the hour."

"I doubt we have that much time. I must see him." Reluctantly, Fenella bent to remove her dirk. A small gossamer pouch fell out of the sheath and landed on the woven thatch floor.

Talfryn's brow crinkled. "What is that?"

Fenella scooped up the tiny sack. "'Tis so light, I forgot aboot it. Remember? Milena gave me the pouch as we were leaving the Romani camp. She said, 'Wear it around the neck to reduce fever.'" *How could she know I would need this? Is she a seer?*

Puffball scampered off the bed, snatched the pouch, and ran back to Disa. "Try it!"

Beacon shivered violently.

"He is worsening!" Disa said.

Beacon's teeth chattered with such force, Fenella feared they would break.

"More blankets and liquids," Hilde said.

Talfryn conjured thick quilts and placed them over his father.

"Liquids, aye, but nay more tea." Fenella remembered how Edward had brought down her fever after Argonshire's men had stolen her. She faced Talfryn. "Now he needs fresh mint and clean spring water."

Without answering, he disappeared.

Disa placed the gossamer pouch at the base of Beacon's throat. As soon as the fabric touched the fay's fiery skin, the pouch became bright orange.

"Fascinating," Hilde said. "Magic?"

"Aye," Fenella said.

"Good magic for humans may be bad magic for fairies," Hilde warned.

"True." Fenella worried her thumb across the scar on her palm.

Disa removed Beacon's ring and returned it to his limp finger. "Come back to me, my love," she whispered.

Beacon angled his head towards his wife.

Open your eyes, Fenella willed.

Disa pulled an exquisite, embroidered quilt over Beacon's shoulders, then mopped his brow with rosemary water.

Beacon's shivering eased.

Talfryn returned with the tangy-smelling mint leaves steeped in fresh spring water. Disa squeezed a few drops between Beacon's cracked lips. After a few moments, the Romani pouch faded to a cool melon shade.

"'Tis working!" Fenella pressed her palms together.

"The sun is about to set," Talfryn said. "We must leave afore Mab comes for you."

"Nay. Not yet," Disa protested.

"Mab? Then I must depart as well." Hilde glanced towards the tunnel.

"I will not leave him alone," Disa insisted.

"I will stay with Beacon," Puffball volunteered.

"Reopen the tunnel as soon as it is safe," Hilde said. "I will return." She squeezed Disa's hand. "Ye did better and lasted longer than those who came afore ye. Ye almost succeeded in bringing the elixir back to the mage. Grambolyn would be proud."

Beacon's moan covered Hilde's yelp as the dwarf transformed back into a wolf and dashed into the tunnel, dragging her mantle.

The pouch shifted to a soft, healing shade of blue.

Talfryn placed the thunderstone necklace over his father's head.

Beacon's eyes fluttered opened. "Take your leave, my love," Beacon told Disa, his voice little more than a hoarse whisper. "We will meet again."

Disa bent and pressed her cheek to his. Puffball mewed. Fenella crept to the door of the chamber to give the family a bit of privacy.

"I kept my promise, dear wife," Beacon whispered. "I learned the names of Mab's human slaves—"

Names of slaves? Those three human children have names, but I have never asked who they are or where they came from. When we returned the potion, Mab… Och! Worried about Mab's decree, Fenella, retrieved her dirk, then fled to the nearest room with a window. Mere moments passed afore Talfryn and Disa joined her.

"Father wishes to say farewell to you."

"We haven't time."

"Please?"

Disa extended her hand. "Go. I will keep safe your weapon."

Fenella gave her the dirk, then returned to Beacon's chamber.

The tunnel had disappeared, and the floating bed was again aloft. It rocked gently as Beacon lifted his long-fingered hand for her to clasp.

"Closer," Beacon rasped.

Fenella bent so her ear was close to his mouth. As she did, Beacon pressed the gossamer pouch into her hand.

"If Mab discovers you have found a partial cure for Cold Iron poisoning, she will enslave you along with the girl who gave you this magic."

Fenella drew back. The truth of his words sent a chill shivering through her. She glanced at the bedside table. *He vanished the tea or rendered the bowl invisible.*

Beacon strangled a cough. Seeing he had more to say, Fenella moved closer. The floating bed swayed.

"Bury the pouch somewhere safe. Send your cousin to protect the Romani."

"But Sorley is just a lad."

Beacon shook his head, then fought back another cough.

"My recovery must appear slow and uncertain to lessen Mab's suspicion. Tell Connell to take good care of my wife. Now hasten away afore Mab grabs you!"

"Farewell." On impulse, Fenella kissed the fairy's clammy cheek, then fled the chamber.

In the antechamber, Disa handed her the dirk. Mindful that Mab's spies could be anything, even the spider spinning a web in the corner, Fenella maneuvered her fairy cloak so it covered her hand as she returned the blade and the pouch to her boot.

"Quick! Join hands!" Talfryn said.

"Farewell!" Puffball called from the bed.

As Saharan scents assaulted Fenella and the sun vanished, a painful prickle rippled across her scalp. The stings spiraled from her head to her neck, then gained speed as they nettled down her body in a race to her toes. Her scream shattered into mute shards as she disintegrated into milky smoke.

CHAPTER FIFTY-TWO
FAREWELLS

Thorfell, Scottish Highlands

With Puffball's farewell fading in Disa's ears, the shield-maiden doubled over in pain as she shattered like a smashed crystal vial. Fragmented, she entered a disorienting dark void where explosions of white light bloomed, then vanished. Darkness swallowed her screams. At last, the sharp slivers of her being reformed with stabbing pains so fierce, they catapulted her out of the Netherworld. She materialized on a chilly heath.

Beside her, an unnatural disturbance of air expelled the thaness, who tilted, then sidestepped to regain her balance. Disa clasped Fenella's elbow to steady her. "Goddess, protect us!"

"Bluidy—" Fenella found her footing. "May it be so."

Disa glared at Talfryn, who with an expression of great apprehension stood within arm's length, clutching his spiky hair. Softly lit by a glow orb hovering above his head, his features relaxed as he lowered his arms, then reached for her hand.

"I feared I had killed you both! When I smelled Mab approaching, I had to act quickly." With a flick of his finger, Talfryn directed the glow orb to pool light over them. "You are so pale. Are either of you hurt?"

Já. Everywhere. Disa exchanged a sideways glance with the thaness. The wan redhead straightened her spine and flattened her lips together.

"We are here. That is what matters." Taking a deep breath, Disa inhaled briny sea scents as she peered into the darkness. The moon, still full and bright, shined upon a stone tower house perched upon a nearby rise. Two men, night watchmen, Disa surmised, walked atop the high stone wall that concealed the lower half of the fortress. She drew her cloak tightly against the cold. "Did Talfryn's magic send us to the right place?" she asked Fenella. "Have we reached your uncle's stronghold?"

"Aye." The thaness stepped into the moonlight, and with three sweeping arm movements, appeared to signal the guards on the ramparts. One man hastened from his post, disappearing from view. "Connell's deerhounds will soon be upon us, followed by guards bearing Cold Iron."

Talfryn vanished the orb. "Then this is farewell." His voice cracked. "Mother, I am sorry you lost the Aeternitas Potion to Mab after safeguarding it for so long."

With a fervent hand gesture, Disa motioned them closer and whispered, "Mab does not have the potion, nor do I. When you cut the seam with Beacon's thunderstone, I noticed the thread was a slightly different color."

Talfryn gawked. "When had you last examined the cloak?"

"The day you found me, but I was in the dark farmhouse and merely felt for the vial. I thought it strange when it warmed my fingers through the cloth. It had always felt cool. I last saw the original sewing, a few hours afore Torolf returned."

"Torolf is the druid farmer who gave her shelter," Talfryn explained to Fenella. He clapped his hands to his head. "Merlin!"

The thaness creased her brow. "What part does Merlin play in this?"

"When Torolf departed with Signy, he glanced back. I wondered if my eyes deceived, but in that instant, Merlin's goat-bearded face emerged from Torolf's."

I slept across the hut's fire from a great sorcerer! Disa clutched Talfryn's arm, suddenly weak in the knees. "I believed it odd that Torolf had returned after taking his family to safety."

Talfryn appeared stricken. "Merlin took Signy."

"Surely, Merlin withdrew from Torolf once he had the potion," Fenella suggested. "Then your friend returned to himself."

Disa flashed on the young goat-bearded druid who had appeared to her through Ragna so long ago. *Goat-bearded sorcerer!*

"Wouldn't Torolf wonder what he was doing back on the island with Signy?" Talfryn asked.

"They are all druids," Disa reasoned. "Mayhap Merlin aided Torolf and Idony when they fled the island, then asked a favor in return." *He aided me when I needed guidance. Foretold my allies. And unknowing, I brought him the Aeternitas potion. But what does Merlin want with a potion that holds the key to uncovering the truth about the Last Fairy King?*

Deerhounds' gruff barks sounded from behind the wall.

"Merlin would have taken Signy to Avalon if he sensed magic in her," Fenella said.

We are but pawns unaware of what Merlin plots. Disa tightened her grip on Talfryn's arm. "You must search for Signy as soon as Beacon has recovered. Ensure her safety."

"I vow, I will."

Satisfied, Disa released her hold.

"Fenella," Talfryn said, switching his attention to the thaness, "you told me that in your time, Merlin and the Lady of the Lake are just characters in stories written long ago. Did any of those stories mention Signy or the Aeternitas Potion?"

"Neither was mentioned. But when King Arthur was mortally wounded, Morgan le Fay brought him to Avalon where he is rumored to be in an everlasting sleep 'til his country needs him again. What if the stories are true, and the Aeternitas Potion was used to place Arthur in a dream world between life and death?"

"By the gods!" Disa pressed her fingertips to her lips. *I dared not imagine such powerful magic.*

The iron portcullis protecting Connell's stronghold creaked upwards and two long-legged deerhounds raced towards them.

Fenella wound the edge of her mantle around her hand. "If Merlin switched the vials—"

"Mab won't be able to cover up any more murders," Talfryn said.

"Or eternally enslave human children." Disa's eyes shined with victory.

Connell and three men holding torches rode out from the stronghold.

"Merlin is no friend to Mab," Fenella remembered out loud. "Whatever magic he poured into the vial…"

"Might harm or even kill whoever drinks it." Talfryn shifted his gaze from the thaness. "Mother, you will be blamed. We will all be answerable."

Beacon! Disa stiffened her legs to prevent them from buckling.

"Talfryn, you must warn Vendel and Hilde." Fenella's air of authority gave Disa hope. "Then keep your distance, lest Mab concludes you conspired against her."

"I agree." On motherly impulse, Disa enveloped her son in a brief, forceful embrace. "Be ever on guard and hasten your father to a safe place beyond Mab's reach if Merlin's magic is unleashed."

"The Callanish Stones," he whispered in her ear. "Whatever year the switch is discovered. I vow."

Disa nodded, then released him.

"Fenella." Talfryn cleared his throat. "When I magicked us here, I aimed for us to arrive on the same day you escaped the Blackwood stronghold. If it worked—"

"I will have a wide advantage over the men searching for me. Thank you."

"It was the least I could do after... I am so sorry." He bowed his head.

"I would have done the same to save my mother." Fenella lowered her gaze and turned towards the approaching riders.

Talfryn closed the gap between he and the thaness. "Why can't you return to Blackwood?" he asked in a tone so low, Disa barely overheard.

The moonlight haloed Fenella's riot of red hair as her shoulders slumped. Then, facing him once more, she said, "Because his father intends to murder me as soon as Edward and I marry."

As Disa watched, Talfryn's mouth fell open. "Is Blackwood party to this?"

"Nay. You saw his face when he ran after me." Her sweeping gaze included Disa. "'Til I am of age or marry, I remain Argonshire's ward. His prisoner. I have two choices. Marry Edward Blackwood and hope he loves me enough to protect me from his father. Or—"

"Let me protect you." Talfryn clasped her shoulders. Then eyeing the fast-approaching riders and dogs, he released her and stepped back.

"You canna protect me." Fenella clutched her mantle. Her eyes conveyed strength of conviction. "Only *I* can truly protect myself."

"Já. 'Tis the way of the world," Disa said. "Will you train to become a shield-maiden?"

"Nay. I would be one sword against many." Fenella's expression hardened.

Talfryn appeared more worried. "Your determined look reminds me of the Lady of the Lake and her handmaiden, Glynis."

Fenella snorted. "How fitting."

Disa, recognizing something of herself in her new ally, clasped the thaness's hand and stared into her eyes. "You have determined your fate."

"Argonshire fears magic."

"Magic cannot be denied, nor can Fate be fooled." Disa said, certain her resolute expression matched the thaness's. "You will learn to control your magic. Master it."

"Aye." Fenella raised her chin. "I must become what Fate has decreed—a sorceress."

Chapter Fifty-Three
Lily

Belle Flor, The Noble House of Storms
The Realm of Air, Fairy

Talfryn alighted softly on pale pink flagstones, not at all where he had intended. *Why have I landed in the maze garden, a place for pondering weighty decisions?* Moonlight bathed the sculpted hedges and flowers flanking the path. He sensed hidden baby flower fairies sleeping among the night-blooming jasmine and snowy moonflowers. One nestled within the evening primroses cooed.

He trenched his brow. His forehead ached. Questions about Fenella, his parents, Lily, and the future collided inside his head. Talfryn raked his fingers through his hair. *I have nay time for the maze nor contemplation. Once I ascertain Father is safe, I must find Lily and give her the seeds.* As he directed his thoughts, the sound of trickling water reached him. *The fountain!* He followed the splashing noise to the clearing at the end of the maze. Afore him loomed the glorious moonlit spires of Belle Flor.

The glamour has lifted. He flashed on when his father had first brought Fenella and Sorley to Belle Flor. Not wanting to beguile the humans, Beacon had glamoured the towering, cloud-bright castle so it appeared to be a simple cottage with fanciful gardens. The glamour had reappeared earlier today when Father was near death.

He rubbed his palm against the sudden tightness in his chest. Squaring his shoulders, he rounded the stone fountain. Glowing violet spheres bobbing on the breeze trailed light across the dark reflection pools. Talfryn glanced up at the clouds topping the entrance arcade. *Nay Trooping Fairies guard Father from the air.* The tightness spread to his stomach. He pulled on the crystal dragonfly door handle and slipped inside.

Talfryn paused in the domed side entry where blue-gray walls shimmered with embedded lights and crystals. Familiar voices, pitched quiet, drew him towards the blizzard-white antechamber adjoining his father's sleeping chamber, where he had meant to land.

"Lily?"

She met him in the doorway, her glamoured wings fluttering. "You're back!" She stood on tiptoe and bestowed a quick, butterfly-soft kiss on his cheek. "My parents are with your father and Puffball."

"I feared no one cared."

Lily linked her arm though his. "Of course, we care."

"Talfryn." Lily's mother, Willow Weep, head of the Noble House of Boughs, greeted him. "We came to watch over Beacon in your absence."

"Mab arrived seconds afore us," Lily said in a rush.

A smug expression lit her father's face. "Our dear queen was approaching Beacon when we materialized," Ash said.

"Surprised by their presence, our haughty queen disappeared as fast as she had appeared!" Puffball recounted. The pixie leapt off the bed and scampered to Talfryn's boot.

"Good riddance." Talfryn wrinkled his nose at the desert scents still lingering in the chamber. "I am grateful to you all."

"An assault on one noble, is an assault on us all. The Noble House of Boughs remains steadfast in our alliance with the Noble House of Storms." Willow Weep stared down at Beacon who had fallen into a fevered sleep. "It is a miracle he survived."

"Aye." *Thanks to Fenella and the Roma magic.* "Could you keep watch a few more minutes? Lily, there is something I must show you."

Ash and Willow Weep exchanged a quick glance. "Keep your wits about you." Ash warned. "Merrybelle's murderer has yet to be discovered."

"We will be quick and careful." Talfryn faced Lily. "This way." He led her into the herb garden where, earlier, Fenella had collected leaves for the curative tea. "Do you trust me?"

Lily's delicate brow creased. "Of course. What is this about?"

He clasped her hands. "You will see."

Talfryn kept his fingers firmly laced through Lily's as the two of them materialized on the edge of the burnt forest. The acrid scent of charred wood and spent dark spells assailed them. Lily's eager expression vanished. Her nostrils flared, and her eyes welled. She tried to pull from his clasp, but Talfryn held fast.

"Why did you bring me here?" The new glamour Lily had cast to cover her welts and burns wavered. "Release me so I may return to my parents."

The hurt in her voice clawed his heart. Talfryn eased his grip. *This was a mistake. It is too soon.* "I can explain. Please. Hear me out."

She glared at him. "I grant you one minute."

"My search for the Aeternitas Potion brought me to a northern island." Talfryn released her.

Lily raised her milky hand to cover her nose and mouth from the grit and stench. Under the full moon, the ashes resembled snowdrifts.

Whiffing the mordant air, Talfryn continued in a hoarse voice. "Norse raiders were torching every dwelling and tree." The heat of the fires rushed back to him, as did Signy's terrified face. Sweat prickled his brow, and he swallowed hard.

"I thought of you when the Lady of the Last Forest gave me this." He retrieved the sack of seeds from his pocket and handed her the bulging pouch. "It contains alder, hazel, birch, and ash seeds. The Lady bade me plant the seeds where fairies can protect them. I could think of no better protector than you, Lily. Nor a better place to replant than here, where your family lost their home."

Blinking back tears, Lily pressed the sack to her heart. "The last forest? Truly?"

"Aye. Truly."

Eyes brimming, Lily jutted her pointy chin towards the charred remains of a nearby tree. "Merrybelle and I used to swing from a vine rope tied to a high branch on that pine. Its tree spirit was always welcoming to us. I hope she escaped in time."

Talfryn's throat constricted as he remembered the wildwood's terrified tree spirits. He glanced about the scorched woods. *I am breathing in ash—Merrybelle's ashes.* Suddenly unable to take a deep breath, he felt as though he were suffocating. *Any surviving trees were too far away for the tree spirit to have leapt.*

Lily tilted back her head. Talfryn realized she was staring at the empty space where her family's treetop home had perched.

"Merrybelle claimed she could see Avalon from her chamber window."

"Through the mists?"

Lily nodded. "My sister admired the Lady of the Lake. When we were younger, she often imagined what it would be like to be one of Viviane's handmaidens. But until the night afore she died, she would smile in an enigmatic way, and hint she would never leave Fairy because of her secret admirer. Merrybelle never told me his name."

She couldn't tell you because her admirer was the Prince Consort. "Did something happen the night afore she was killed?"

"I do not wish to speak ill of my sister."

"If you know something, please tell me. It may be a clue to who killed her."

Frowning, Lily twisted the tie on the seed pouch. Moments passed, then she said, "Merrybelle confessed she had spiked her admirer's drink with a truth-telling potion. She had only wished to know if he loved her more than the fairy thought to be his true love. But instead, the drug prompted him to confide a disturbing secret."

Jolted, he forced a calm tone. "What secret?"

A tear fled down Lily's pale cheek. "Merrybelle refused to say. But afterward, she jumped at every sound, and kept glancing over her shoulder, as though fearful her admirer would realize he had told her something he shouldn't have."

He smoothed the tear from her cheek. "If he thinks you know the secret, then you are in danger, too."

Lily nodded. "My fear as well."

"You and your parents must continue to live at Belle Flor. We'll increase the protective wards."

"Thank you." Her glance returned to the pine that had once held a vine swing. "Merrybelle was nine years older than me. She fancied herself a secret keeper, a collector and protector of Fairy secrets. Once, she teased me that she knew a secret about me, but she refused to reveal it. When I asked my parents about it, they glared at Merrybelle, and she never mentioned it again."

Talfryn shook his head. "I would have gone mad with wondering."

"If not for you, I might have."

"Me?"

"I would watch for you on the path. Merrybelle and I would invent stories about the Norse. She always cast me as a shield-maiden or queen."

"Excellent choices. You are fiercely loyal and remarkably brave."

"Few people notice." She stared at the ash drifts half covering the forest trail. "When the dark fire erupted, the tree spirits' screams woke me. I ran to my window, looked down, and saw Merrybelle on the path." Lily hugged her thin arms to her waist. "I think she was meeting someone."

Brindleburr. "Did you see who?"

Lily shook her head. A lock of her lavender hair fell across one eye. "Merrybelle was loyal and kept secrets well. I fear her own secret may have cost my sister her life."

He squeezed her willowy arms, then swept the hair from her eyes. The fevered heat of her skin beneath his fingertips tore his heart. "I am so sorry."

"Thank you." She stared into the distance, past the charred trees and ash to where moonlight illuminated healthy trees, green and filled with life.

"I can save the seeds 'til you are ready. Or you could use magic to create a new forest. 'Twould be easier and faster," Talfryn suggested.

Lily sniffed, then shook her head. "The Lady requested the seeds be planted, so I shall plant them." Her expression brightened. "Imagine, watching a forest grow from seeds!"

Talfryn's smile held little glee. "One advantage to being immortal."

Lily's expression flickered. "Merrybelle's death has me questioning my own immortality. If she could be murdered—"

"I promise, nay harm will come to you." He bent and pressed his forehead against hers. "I intend to live long enough to see most of your wildwood grow."

"*All* of it."

"Mayhap, if you keep saving me, my chances will improve. I thank you for your help landing on the moor of Blackwood's Castle instead of plunging to certain death into its loch." He stepped back and shook his head. "Lily, the risks you took."

Her cheeks flushed to the shade of dog rose blossoms. Then her brow furrowed. "Why didn't your mother drink the potion? Didn't she wish to be young forever, like Beacon?"

Talfryn hesitated. He longed to confide that his parents suspected Mab had murdered the Last Fairy King and Gudbrand the Dwarf. And that his mother had promised Grambolyn she would find proof or stop Mab from killing again.

Lily stared up at him. Patient. Concerned.

Talfryn stuffed his parents' secrets into a shadowy corner of his heart. *I guess I, too, am a secret keeper.* "Disa knew that if she consumed any of the potion, there would be nay hope Mab would allow her to see Beacon or me."

"Your mother knew Mab would banish her after your birth, so she stole the potion as barter. Very clever, but she took quite a chance."

"Aye." *More than you may ever know.*

Lily's gaze dropped momentarily to Talfryn's onyx ring. Frowning, she smoothed her fingers down one of her long lavender locks. "When you visit your mother, will you seek out the thaness?"

"Fenella is in terrible danger." *How many paths had the drabardi seen in Fenella's palm? Two? Three?* He plowed his fingers through his spiky hair. "I wish I could protect her."

Lily clasped his arm. "Let humans settle their own affairs. Didn't the thaness's king command she marry Edward Blackwood? Let them be. Beacon needs you here." She gripped harder and pressed her wings together. "I need you."

"Lily, I—" Conflicting loyalties jumbled his thoughts and froze his words. He frowned down at the onyx ring Stanka had given him and flinched. Blackwood stared back at him. Their eyes met in a silent challenge.

"What's wrong?"

A sense of foreboding shivered Talfryn's flesh. "Nothing." He rubbed his palm over the shiny black stone. When he lifted his hand, the image had vanished. Talfryn forced a neutral expression, then slipped the ring into a sealed pocket he quickly conjured inside his tunic.

"Is that ring a gift from the thaness?"

Lily's hurt tone pierced Talfryn's heart. "Nay. A soothsayer I met during my quest gave it to me. She told me it would protect me, but I suspect it may contain a different enchantment. He closed his hands around Lily's as she held the seed pouch. "I am loath to leave you, but I must retrieve Nettle soon. Don't worry if I disappear for a few days. You know how troublesome Nettle can be."

"You would leave Beacon?" Her voice rose in disbelief.

"Not 'til he has regained more strength and I am sure you are safe." He squeezed Lily's hands. "Show your parents the Lady's gift. Mayhap the thought of a new forest will lighten their spirits."

Lily glanced down at the bulging pouch. "Do you think we will ever discover who killed Merrybelle?"

Talfryn leaned in close, in case Mab's spies were about, and whispered in Lily's ear. "I am quite certain. Keep your eyes on the Court." He drew back.

Lily held his gaze for several heartbeats, as though she were peering into the depths of his soul. Her expression wavered between hope and heartache. Then she stood on tiptoe and kissed him softly on the cheek. "Be safe."

She receded a step, then disappeared in a flurry of scorched rose petals.

Talfryn gaped at the spot where Lily had stood. A sad, hollow feeling settled in his chest.

Puffball popped out of Talfryn's cloak pocket, startling him.

"When did you sneak in there?"

"When you were distracted." The pixie scrambled up on his shoulder. "Have you sorted it out yet?"

"What?" he asked, his tone sharp.

"Fenella or Lily?"

"If Fenella becomes a sorceress, the thaness we know will no longer exist."

"She wants to become like Morgan le Fay and the Lady of the Lake?"

"Aye. Argonshire schemes to murder Fenella once she weds Blackwood."

Puffball hissed air through her jagged teeth. "If Fenella succeeds, she may lose her chance at true love with Blackwood—or you." The pixie frowned. "But she'll save herself."

"Will she? Magic always exacts a price."

"What matter the price?" Puffball grabbed a spike of Talfryn's hair and tugged. "Your parents paid a high price for love, but I doubt either would have acted differently."

"True."

"You and Edward Blackwood should join forces to aid Fenella."

Talfryn considered the onyx ring. "The Blackwoods believe magic is a taint to be eradicated."

"When we were outside Fenella's castle, it didn't sound like Edward Blackwood shared that belief."

Talfryn scowled in disbelief.

"Choose your course soon, or risk losing Fenella *and* Lily," Puffball warned.

"I fear I am already too late." *Two paths,* the drabardi had told him. He unfurled his hand. Seeing the two lines remained, his breath caught.

There is still time.

Chapter Fifty-Four
Family

Thorfell, Scottish Highlands

Mab bound Beacon to a rowan and cuffed him with Cold Iron! Disa inwardly shuddered as her thoughts raged. *By the gods! I should never have left him. Should he worsen…* Standing in a shaft of morning sunlight in the Scot's Great Hall, she closed her eyes and tried to focus on the thaness's words.

"Disa?" Fenella's tone was soft and compassionate, as if the thaness could read the shield-maiden's heart. The redhead, her cousin Sorley, and their uncle Connell were grouped beside her at a trestle table. Disa shifted, and light from the bay window spilled onto the map spread afore them. The rendering depicted the vast Thorburn and Rose lands that constituted Fenella's thanedom.

She fingered the crystal sphere tucked inside the pocket of the léine Connell Thorburn had procured for her. *How odd to be dressed thus. My movements are somewhat restricted,* she thought as she released the crystal and wiped her hands down the front of her long tunic. The léine sparked a memory of her sister, Bree, twirling in a green meadow on their parents' small farm. The morning sun had haloed Bree's wheat-blond hair, and her léine had ridden up past her ankles when she lifted her arms.

Bree. Disa's throat thickened, and she swallowed hard, remembering the slavers' block. What had Ragna told her so long ago in the diviner's cave? *"Always remember. You were not born a slave."* And Talfryn had echoed her counsel. *Mother, you were not born a slave.*

"You must be exhausted," the thaness said.

"Aye. My thoughts were with my family."

Fenella and Connell exchanged a worried glance.

"We will keep you safe while you await Beacon," Connell said.

"Thank you for sheltering me and offering me protection. But I will find work and a hut in the village." She raised her chin. "I have reached a decision. My debt to Grambolyn has been paid. I do not need his permission to be a freewoman. I was born free. I *am* free."

"Aye!" Fenella embraced her. Disa stiffened at the unfamiliar gesture, then relaxed and returned the young woman's tight clasp afore stepping back.

"Beacon requested I protect you. Please allow me that honor, at least 'til he can join you," Connell implored.

"You are too kind." Disa wrapped her wool mantle more closely around her torso. *If...when Beacon recovers, he will find the Highlands far too cold and the swarms of tiny bugs vexing.* "He will come. As soon as he is able."

"Absolutely," Fenella said. "And Vendel will find a way to return your sword and shield. Surely we will hear from him soon."

"In the meantime, you are welcome to any weapon in my armory," Connell said.

"Your generosity overwhelms me."

"Cousin, you were explaining the map," Sorley prompted.

"Aye," Lady Fenella turned back to the parchment on the table. "'Tis vital you understand the boundaries."

The Highlanders' accent challenged her ears as she followed attentively what the younger woman was saying. To be certain, Disa confirmed, "If I stay within these borders, I will abide by our agreement with Mab?"

"Aye." Fenella traced her finger along a shaded double line on the map. "As you can see, three of the strongholds are along the river."

Connell pointed. "And we are here, by the firth."

Lady Fenella's features hardened, mayhap with the resentment and rage Disa suspected brewed beneath her surface. "You will nae be safe there." The thaness stabbed her finger at the drawing representing her castle. "Nae 'til I retake my thanedom."

"But should you ever need to flee, the Roses will welcome and protect you." Sorley indicated the two strongholds closest to Thorburn Castle, one to the southwest, the other adjacent on the east.

"But only Connell and Sorley know aboot Beacon and Talfryn," Fenella warned.

"I am indebted to you all," Disa said.

"If any debt is owed, it is the one I owe Beacon," Connell said. "You do me a great favor and honor by being my guest. You are welcome for as long as you wish."

"Thank you." Disa slipped her hand into the pocket containing the crystal sphere. "I accept your hospitality and protection… for now."

Fenella turned towards her uncle. "Have you sighted the Romani recently?"

"Aye. They have made camp there." Connell traced his finger along a sparse forest at the northern edge of the thanedom. He pointed at a small glen nestled against the tree line Disa judged to be within a half-day's ride.

"They have moved north since I last saw them." Fenella pressed her knuckles to her lips.

Fatigue from last night's charged discussion shadowed Connell's eyes. "We should consider more options."

"This is the only way." Fenella caught Sorley's gaze. "We will leave within the hour and stay ahead of Lord Argonshire's men. After I speak with Stanka, we may have to part ways."

Disa rolled the crystal sphere between her forefinger and thumb. "You seek the Roma's counsel on becoming a mage?"

"A sorceress," the thaness corrected.

"Your betrothed will be searching for you," Connell said. "Mr. Blackwood sought the *drabardi's* counsel the last time you vanished. He may do so again."

"Aye."

Disa saw the blush flame across the thaness's cheeks. She changed subjects. "The Romani have my son's other pixie?"

"I believe they still do," Fenella answered, "unless Talfryn has already returned for Nettle."

"Then I will journey with you," Disa announced. "I wish to meet the soothsayer who led you to me."

"'Tis too dangerous," Fenella protested. "Mab specifically vowed the *Blessed* Court wouldna harm you. But she offered nay protection from the *Unblessed* Court."

Redcaps. Snatchers... Disa rubbed the fine blond hairs rising on her forearm.

One of the deerhounds sleeping by the hearth raised its grizzled head. Disa glanced at Connell and noted the determined set to the Scot's jaw.

Fenella added, "Uncle, you must stay and protect Disa."

Connell crossed his arms. "And who will protect you from Queen Mab or Argonshire's men? A few of my men and I will accompany you. The rest will remain with Disa."

"A contingent will draw unwanted attention. Sorley and I can move more quickly on our own."

The deerhounds suddenly skittered across the great hall, barking excitedly as they headed for the door leading to the outdoor stairs.

Disa tensed and glanced out the bay window. *Nay warships.*

Fenella and Sorley hastened to the arrow slit overlooking the courtyard and gate.

The deerhounds lunged at the door, scratching the wood with their nails.

"What disturbs them?" Disa reached for her sword, forgetting she had left it on the headland.

"Nay, 'what,'" Fenella answered. "Who."

Sorley, with what appeared to be cautious hope shining in his eyes, said, "Mayhap we willna be traveling alone after all."

Disa joined Fenella at the arrow slit and peered outside. A dog taller than Connell's huge deerhounds approached the gate. "'Tis a wolfhound."

"Nae ordinary wolfhound." Connell crossed the hall with a swiftness of foot. "Careful, lads," he warned the deerhounds as he opened the door. The dogs bolted.

"I am confused," Disa admitted, her accent thickening. "If the wolfhound is not ordinary, then what is it?"

She watched as Fenella threw Sorley a questioning look. When the boy's furtive glance flicked to her, Disa crossed her arms expectantly.

Sorley seemed to hesitate afore he looked to the thaness once more and nodded.

"Family," Fenella explained as Sorley, appearing anxious and relieved, shape-shifted into a russet-furred wolfhound as he bounded out the door to greet his father.

EPILOGUE

Disa stood afore the bay window in Connell's Great Hall and stared out at the blue-gray firth. Imagining the waters flowed to Fairy and the Romani camp, she cast silent wishes that Beacon would recover quickly, that Talfryn and the thaness would be safe, and that Bree, wherever her soul may rest, was at peace. Mayhap her wishes would reach Fensalir, the wetland halls of the goddesses Frigg and Hlín.

She fingered the heart-shaped stone Bree had given her the day they had encountered Hlín near the fairy mound. Disa remembered how the fairy cat Muirgen had yowled when Bree had said her soul protector was a white cat with tufted ears. Mayhap the affectionate tabby thought *he* was Bree's protector. The cat's overly large, round fay eyes, normally gleaming with delight and mischief, still perplexed her.

"I thought you would be asleep by now," Connell broke her pondering when he spoke from the doorway. "Does your bedchamber nae suit you? I will find you another."

Disa pocketed the stone as she twisted towards him. "I thank you, but the difficulty is with me, not the bedchamber."

"Ah." Connell crossed the hall and joined her by the window. Beneath a sky streaked orange and dusty pink, a single boat bobbed in the water. "'Tis late for fishing."

"Aye. Were it not so close to the solstice, darkness would have fallen hours ago."

They stood in silence, watching the glittering firth mirror the darkening sky.

Connell sighed. "It never gets easier."

Disa glanced his way. "What?"

"Saying farewell."

"You are thinking of this morning."

"Aye. Were you nae musing over your farewells?" Connell asked.

She glanced down at her wedding ring. "Beacon and I never say farewell, lest Fate overhears us and removes all hope of our reuniting. Instead, he told me of a vow he had fulfilled."

"A vow?"

"Long ago, I had accused my husband of not caring about Mab's human slaves. *Children* unable to flee Fairy 'til Mab's binding spells expire!" She shook her head.

"Did Beacon free them?"

"Nay. 'Tis impossible 'til the magic fades or the fairy that cast the spell dies. But he had vowed to learn their names."

"And did he?"

"Aye. Baird, Oscar, and a redheaded girl who curiously, has a male name. Esme."

Connell coughed violently. "Did you say, Esme?"

"Aye. Have you heard of such a girl?"

"Mother claimed Esme had been stolen by fairies, but no one believed her." Connell seemed lost in distant thought afore he looked at Disa and asked with earnest, "Did you see the lass with your own eyes?"

Disa nodded. She shifted from one foot to the other, remembering the two dark haired boys, also enslaved. *Were they the children Ragna had mentioned, the boys who had failed Grambolyn's test and been traded to a fairy?*

Urgency tinged Connell's words. "How old did she appear?"

"Seven winters, mayhap eight."

"By the stone!" He paced afore the window. "At first light, I shall send Mother a message. Mother was barely ten when Esme disappeared. Mayhap she still has the sketches she made of her sister. Sadly, Mother is nae a talented artist, but still..." He strode towards the door, then abruptly halted and faced her. "'Twill be dark soon. I will send a maid to escort you to your chamber."

"I thank you." Disa watched him hurry from the hall. *Could the red-haired slave child be a member of Fenella's family? I pray he will be able to rescue her.* She flashed on Bree and a familiar ache shackled her heart.

Deep in thought, she gazed out the small panes that made up the large hall's window. The lone boat, swaying on the waves, appeared black against the setting sun and sparkling water. Memories coasted into her thoughts. Rocking in her small vessel on her own solitary voyage. Searching the waters of the unnamed sea for the seer's cave. *Seems so long ago now.*

The sudden reflection of a brilliant, blue light cast from behind her, its glow filling the glass beyond her own fractured, reflected image and obscuring the view, wrenched her from her reverie. Disa spun and raised her hand to block the glare. "Hlín."

"You remember me." The armored goddess sounded pleased. Her gaze slid over Disa's borrowed feminine clothing. "A shield-maiden without her weapons is a woman half lost."

"'Tis true." Disa glanced down at her hip where her scabbard once hung. "I feel as though pieces of me are missing. But the cause was just."

"Your husband's life."

"Aye."

"My queen and I agree." Hlín faced her palms skyward. A familiar circular shield appeared in the goddess's hands. Upon it were Disa's helmet, scabbard, and sword. "Reclaim what is yours."

"I am humbled and deeply grateful." Disa accepted her accouterments, awed by the goddess's unexpected gifts. She placed the weighty shield on the nearest trestle table, then strapped the scabbard to her waist. Lifting the sword, she noticed a storm cloud stone with white rings had been added to the hilt. Disa flashed on the thousands of stones on that beach…"Is this—"

"The wishing stone you tossed into the waves the day you saved your sister from drowning. Aye. Your sister's magic retrieved it."

"My sister…" Disa could barely breathe. "Bree lives?"

Golden rays beamed from Hlín's palms as she gestured towards Disa's sword. "Behold your blade."

Holding the sword crosswise, Disa balanced the tip on her left palm. The blade clouded as though covered in mist. An odd sensation tugged at Disa's heart, and she felt herself pulled into the cool, damp vapors.

In a dreamlike state, she floated through the mist, landing lightly within a coracle. Swift waves splashed the sides of the small circular boat. The mist parted, revealing an azure sky and an island close ahead. Disa's chest thrummed as palpable magic propelled the coracle towards a young woman and a tabby cat waiting on the shore. Sunlight glinted off burnished strands threaded through the teen's wheat-blond hair.

Rising to her feet, Disa nearly tipped the coracle.

"You are finally here!" The young woman hoisted her long blue tunic up to her knees and ran into the water.

As she drew closer, Disa's doubts faded. *'Tis really you.* "Bree!"

When the coracle slowed in the shallows, Bree grabbed the boat to steady it. "You found me!"

"Where are we? How did you find your way here?" She tried to climb out, to embrace Bree, but an invisible force rooted her to the hull.

"I escaped the slave pen, then Muirgen saved me." Bree glanced back at the tabby on the shore. "We fled through Fairy, evading the Courts and all manner of danger. You would wonder at the sights I saw! Finally, we reached safety here…in Avalon."

"Avalon," Disa repeated softly. She lifted her gaze to the clutch of stone beehive dwellings perched high on a verdant hill.

"Bree refused to leave you."

Disa followed the soft voice to a slender, shirtless male who stole his way down the shore towards them. His fay green eyes twinkled. With a catlike motion, he smoothed his ginger hair, swirled with black stripes, towards his face.

"Muirgen?"

The fairy purred, then twisted to reveal his striped tail.

"By the gods!" Disa felt light-headed.

Bree clasped Disa's hands and stared into her eyes with the intensity of someone who has only moments left. "Please know Muirgen tried to save you, too. He hid me, then went back for you, but you had already been sold and spirited away."

"I was taken to a seer who dwells within an island cave in the unnamed sea. Then I journeyed with my master to a hidden village far from Dubh Linn, where all the inhabitants were dwarfs. You never would have found me. Know that I never gave up hope I would find you again."

"My dear sister, we have found each other." Bree's smile reminded Disa of sunshine, mermaid tears, and twirling in fields of flowers. Years of worry and guilt lifted away, leaving Disa giddy and unimaginably light.

"My love," Muirgen splayed his hand across Bree's back, "the enchantment—"

"Is about to end. The magic had to be short-lived to be powerful enough to urgently draw you here." Bree released Disa's hands and leaned in close, her expression newly grim, as she continued in a hushed whisper. "Muirgen has spies in both fairy Courts. We must warn you—"

A rushing sound filled Disa's ears, drowning out Bree's words. Pain knifed her back, behind her heart. "Nay!" She clutched the rim of the coracle. Bree reached for her, but the magic yanked Disa backwards through the thickening mist. Losing sight of her sister and Muirgen, she wailed, long and silently 'til, abruptly, Disa found herself alone in the Great Hall.

Seeing her own eyes stare back at her in the blade, Disa lost her grip and the sword tumbled from her hands. Lightning quick, she caught it midair. As she lifted the weapon, Disa angled the blade towards the fading light, desperate in her hope to catch a final glimpse of Bree. But the dwarf-forged blade reflected only the sun's fleeting rays.

Unearthly cold ghosted against her nape. Heart thundering, Disa raised the sword as though it were a mirror. Behind her, its image captured in the blade, stood a brown bear, haloed in gold. The *fylgja* inclined its great head toward the bay window. Disa twisted, tracking the apparition's gaze, and was blinded by a burst of orange light launched from the lone boat on the firth.

She raised her hand to shield her eyes. As she squinted out the window, the spirit bear moaned.

Disa's breath hitched. *Not light, fire.*

The flaming ball sped with supernatural speed, growing larger as it neared.

From the doorway, the maid screamed.

"Draw back!" Disa yelled as she dove under a trestle table. The mage's crystal sphere escaped her pocket and rolled out of reach. "Nay!"

The fireball hit with an ear-splitting crack.

Splintered wood, shattered glass, and fearsome flames showered the hall as Disa crawled with her sword towards the magic crystal orb. A large, jagged shard knifed her outstretched arm, pinning her to the floor. Disa hissed in air. Trembling, she released her sword and, with her teeth gritted against the searing pain, wrenched away the shard. "By the gods!" She clasped her arm in a futile attempt to stem blood spurting from the wound.

I must retreat! She reached for her sword, only to recoil when the sudden heat radiating from the iron hilt burned her hand. Blinded by tears, her wounded arm rendered lifeless, she again clasped the bleeding gash, then dug into the timber floor with the elbow of her good arm. Awkwardly, she wriggled backwards without her sword or sphere. Smoke burned her throat as she wormed beneath the trestle table to gather her strength.

Fate offered nay respite. Flames consumed the surrounding tables and rushes. *I must flee!* Disa rolled away from the blaze, then staggered to her feet, coughing, and dizzy from blood loss. Tears caused by the increasing heat and smoke blurred her vision. *Where is the door?*

The fire arced, igniting herbs and reeds strewn on the floor beneath her feet. With a shriek, Disa sidestepped, but the insatiable flames engulfed her léine.

Time slowed as an odd sensation overcame Disa. Her wounded arm hung useless at her side. With her other hand, she beat at the flames consuming her clothes. And with the last breath in her lungs, she screamed.

Disa's spirit bear emerged from the roaring conflagration and rushed towards her, its front paws extended to embrace her life force.

With blissful relief, Disa fled her body and floated beyond pain, impervious to the choking smoke and intense heat. She reached up, her arms and hands glowing, weightless, free of flesh and bone, and clasped the bear's icy paws. *Oh, welcome cold!*

Far, far below, a loud, frantic bell sounded, followed by barking and screams. What appeared at first glance to be long, rippling black cloths, barely discernible against the smoke and deepening darkness, were briefly silhouetted against the blaze. Screeching with murderous anticipation, enormous wings opened wide, talons extended, and with deadly intent, snatcher fairies swarmed through the Great Hall's shattered window and attacked.

She looked down, then glanced away, not wanting to see.

Sudden silence enveloped her. In the quiet, the mage's sphere rattled, then stilled. The mermaid tear whispered, "Ascend. Be free."

Shepherded by her *fylgja*, Disa floated upwards. The Great Hall vanished as night's cloak, inky and cool, wrapped around her. Soon the spirit bear paused their ascent, and emitting low, soothing tongue clicks, drifted from her grasp. Then like a vine wrapped around an invisible column, a vertical spiral of stars materialized. Now a shining light, Disa rose through the center of the whorl. The spirit bear accompanied her, ascending outside the ringlet of stars.

A wondrous radiance formed by millions of stars capped the spiral path. Filled with unimaginable joy, Disa soared towards it, a soul called home.

The End

A Note to Readers

Thank you for reading *The Viking Mist.* I hope you enjoyed it! Reviews are greatly appreciated and can be posted to your social media, Goodreads, or the site where you purchased this book.

Visit my website, **www.AriellaMoon.com**, for news regarding The Two Realms Trilogy, Two Realms Novellas, and my contemporary Young Adult romance series, the Teen Wytche Saga. If you'd like to receive scoops on giveaways and upcoming releases, including *The Solstice Secret,* the exciting conclusion to The Two Realms Trilogy, sign up for my newsletter.

I love to hear from my readers. You can contact me through my website.

May all your days be magical!

~Ariella

About Ariella Moon

A shaman and ordained priestess, Ariella Moon's life is steeped in magic. Naturally, she spent her childhood searching for a magical wardrobe that would transport her to Narnia. Extreme math anxiety, and taller students that mistook her for a leaning post, marred her youth. Despite these horrors, she graduated summa cum laude from the University of California at Davis. She lives a nearly normal life doting on her extraordinary daughter, two shamelessly spoiled dogs, and a media-shy dragon.

Acknowledgments

The creation of *The Viking Mist* has been a long and extraordinary journey. In 2017, I received a thought-provoking critique from my beta reader Carolyn Cole that prompted a lengthy pause while I reconsidered several elements in the original manuscript. I am grateful to Carolyn for inspiring me to dig deeper, shift my focus a few notches, and enrich the characters and relationships in *The Viking Mist.*

My dear friend and talented editor, Barbara Cole, remained undaunted while some characters and plotlines were heaped onto the midden, other plots thickened, and I wove in new twists, some of which will not be revealed until Book 3, *The Solstice Secret.* Thank you Barbara for your commitment to this story!

Appreciation as well to beta reader Mackenzie Eldridge for her line notes, challenging questions, relentless pleas for longer chapters, and for her brilliant suggestion I name Vendel's contraption the Luminositor. Other awesome word suggestions came from London Michaels (equistragons), Suze Datz, and my Facebook followers.

Amanda Matthews designed the gorgeous book cover. Ginger Hanson, JoAnne Hanson, Suze Datz, and Barbara Cole helped with the back-cover blurb, and Suze Datz also proofed the manuscript. My gratitude to them all!

Most of all, a huge thanks to my devoted Two Realms readers and reviewers!

~Ariella

Made in United States
Troutdale, OR
11/03/2026

56844486R00246